FREE Test Taking Tips Video/DVD Offer

To better serve you, we created videos covering test taking tips that we want to give you for FREE. **These videos cover world-class tips that will help you succeed on your test.**

We just ask that you send us feedback about this product. Please let us know what you thought about it—whether good, bad, or indifferent.

To get your **FREE videos**, you can use the QR code below or email freevideos@studyguideteam.com with "Free Videos" in the subject line and the following information in the body of the email:

   a. The title of your product

   b. Your product rating on a scale of 1-5, with 5 being the highest

   c. Your feedback about the product

If you have any questions or concerns, please don't hesitate to contact us at info@studyguideteam.com.

Thank you!

ServSafe Manager Book 2024-2025
2 Practice Tests and ServSafe Study Guide for Food Handler Certification Prep
[2nd Edition]

Lydia Morrison

Copyright © 2024 by TPB Publishing

All rights reserved. No part of this publication may be reproduced, distributed, or transmitted in any form or by any means, including photocopying, recording, or other electronic or mechanical methods, without the prior written permission of the publisher, except in the case of brief quotations embodied in critical reviews and certain other noncommercial uses permitted by copyright law.

Written and edited by TPB Publishing.

TPB Publishing is not associated with or endorsed by any official testing organization. TPB Publishing is a publisher of unofficial educational products. All test and organization names are trademarks of their respective owners. Content in this book is included for utilitarian purposes only and does not constitute an endorsement by TPB Publishing of any particular point of view.

Interested in buying more than 10 copies of our product? Contact us about bulk discounts: bulkorders@studyguideteam.com

ISBN 13: 9781637751923

Table of Contents

Welcome — 1
FREE Videos/DVD OFFER — 1
Quick Overview — 2
Test-Taking Strategies — 3
Audiobook Access — 7
Introduction to the ServSafe Food Protection Manager Exam — 8
Study Prep Plan for the ServSafe Test — 10
Management of Food Safety Practices — 12
Hygiene and Health — 21
Safe Receipt, Storage, Transportation, and Disposal of Food — 31
Safe Preparation and Cooking of Food — 36
Safe Service and Display of Food — 39
Cleanliness and Sanitation — 42
Facilities and Equipment — 48
Practice Test #1 — 52
Answer Explanations #1 — 66
Practice Test #2 — 76
Answer Explanations #2 — 90

Welcome

Dear Reader,

Welcome to your new Test Prep Books study guide! We are pleased that you chose us to help you prepare for your exam. There are many study options to choose from, and we appreciate you choosing us. Studying can be a daunting task, but we have designed a smart, effective study guide to help prepare you for what lies ahead.

Whether you're a parent helping your child learn and grow, a high school student working hard to get into your dream college, or a nursing student studying for a complex exam, we want to help give you the tools you need to succeed. We hope this study guide gives you the skills and the confidence to thrive, and we can't thank you enough for allowing us to be part of your journey.

In an effort to continue to improve our products, we welcome feedback from our customers. We look forward to hearing from you. Suggestions, success stories, and criticisms can all be communicated by emailing us at info@studyguideteam.com.

Sincerely,
Test Prep Books Team

FREE Videos/DVD OFFER

Doing well on your exam requires both knowing the test content and understanding how to use that knowledge to do well on the test. We offer completely FREE test taking tip videos. **These videos cover world-class tips that you can use to succeed on your test.**

To get your **FREE videos**, you can use the QR code below or email freevideos@studyguideteam.com with "Free Videos" in the subject line and the following information in the body of the email:

    a. The title of your product
    b. Your product rating on a scale of 1-5, with 5 being the highest
    c. Your feedback about the product

If you have any questions or concerns, please don't hesitate to contact us at info@studyguideteam.com.

Quick Overview

As you draw closer to taking your exam, effective preparation becomes more and more important. Thankfully, you have this study guide to help you get ready. Use this guide to help keep your studying on track and refer to it often.

This study guide contains several key sections that will help you be successful on your exam. The guide contains tips for what you should do the night before and the day of the test. Also included are test-taking tips. Knowing the right information is not always enough. Many well-prepared test takers struggle with exams. These tips will help equip you to accurately read, assess, and answer test questions.

A large part of the guide is devoted to showing you what content to expect on the exam and to helping you better understand that content. In this guide are practice test questions so that you can see how well you have grasped the content. Then, answer explanations are provided so that you can understand why you missed certain questions.

Don't try to cram the night before you take your exam. This is not a wise strategy for a few reasons. First, your retention of the information will be low. Your time would be better used by reviewing information you already know rather than trying to learn a lot of new information. Second, you will likely become stressed as you try to gain a large amount of knowledge in a short amount of time. Third, you will be depriving yourself of sleep. So be sure to go to bed at a reasonable time the night before. Being well-rested helps you focus and remain calm.

Be sure to eat a substantial breakfast the morning of the exam. If you are taking the exam in the afternoon, be sure to have a good lunch as well. Being hungry is distracting and can make it difficult to focus. You have hopefully spent lots of time preparing for the exam. Don't let an empty stomach get in the way of success!

When travelling to the testing center, leave earlier than needed. That way, you have a buffer in case you experience any delays. This will help you remain calm and will keep you from missing your appointment time at the testing center.

Be sure to pace yourself during the exam. Don't try to rush through the exam. There is no need to risk performing poorly on the exam just so you can leave the testing center early. Allow yourself to use all of the allotted time if needed.

Remain positive while taking the exam even if you feel like you are performing poorly. Thinking about the content you should have mastered will not help you perform better on the exam.

Once the exam is complete, take some time to relax. Even if you feel that you need to take the exam again, you will be well served by some down time before you begin studying again. It's often easier to convince yourself to study if you know that it will come with a reward!

Test-Taking Strategies

1. Predicting the Answer

When you feel confident in your preparation for a multiple-choice test, try predicting the answer before reading the answer choices. This is especially useful on questions that test objective factual knowledge. By predicting the answer before reading the available choices, you eliminate the possibility that you will be distracted or led astray by an incorrect answer choice. You will feel more confident in your selection if you read the question, predict the answer, and then find your prediction among the answer choices. After using this strategy, be sure to still read all of the answer choices carefully and completely. If you feel unprepared, you should not attempt to predict the answers. This would be a waste of time and an opportunity for your mind to wander in the wrong direction.

2. Reading the Whole Question

Too often, test takers scan a multiple-choice question, recognize a few familiar words, and immediately jump to the answer choices. Test authors are aware of this common impatience, and they will sometimes prey upon it. For instance, a test author might subtly turn the question into a negative, or he or she might redirect the focus of the question right at the end. The only way to avoid falling into these traps is to read the entirety of the question carefully before reading the answer choices.

3. Looking for Wrong Answers

Long and complicated multiple-choice questions can be intimidating. One way to simplify a difficult multiple-choice question is to eliminate all of the answer choices that are clearly wrong. In most sets of answers, there will be at least one selection that can be dismissed right away. If the test is administered on paper, the test taker could draw a line through it to indicate that it may be ignored; otherwise, the test taker will have to perform this operation mentally or on scratch paper. In either case, once the obviously incorrect answers have been eliminated, the remaining choices may be considered. Sometimes identifying the clearly wrong answers will give the test taker some information about the correct answer. For instance, if one of the remaining answer choices is a direct opposite of one of the eliminated answer choices, it may well be the correct answer. The opposite of obviously wrong is obviously right! Of course, this is not always the case. Some answers are obviously incorrect simply because they are irrelevant to the question being asked. Still, identifying and eliminating some incorrect answer choices is a good way to simplify a multiple-choice question.

4. Don't Overanalyze

Anxious test takers often overanalyze questions. When you are nervous, your brain will often run wild, causing you to make associations and discover clues that don't actually exist. If you feel that this may be a problem for you, do whatever you can to slow down during the test. Try taking a deep breath or counting to ten. As you read and consider the question, restrict yourself to the particular words used by the author. Avoid thought tangents about what the author *really* meant, or what he or she was *trying* to say. The only things that matter on a multiple-choice test are the words that are actually in the question. You must avoid reading too much into a multiple-choice question, or supposing that the writer meant something other than what he or she wrote.

5. No Need for Panic

It is wise to learn as many strategies as possible before taking a multiple-choice test, but it is likely that you will come across a few questions for which you simply don't know the answer. In this situation, avoid panicking. Because

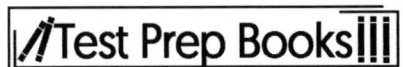

Quick Overview

most multiple-choice tests include dozens of questions, the relative value of a single wrong answer is small. As much as possible, you should compartmentalize each question on a multiple-choice test. In other words, you should not allow your feelings about one question to affect your success on the others. When you find a question that you either don't understand or don't know how to answer, just take a deep breath and do your best. Read the entire question slowly and carefully. Try rephrasing the question a couple of different ways. Then, read all of the answer choices carefully. After eliminating obviously wrong answers, make a selection and move on to the next question.

6. Confusing Answer Choices

When working on a difficult multiple-choice question, there may be a tendency to focus on the answer choices that are the easiest to understand. Many people, whether consciously or not, gravitate to the answer choices that require the least concentration, knowledge, and memory. This is a mistake. When you come across an answer

choice that is confusing, you should give it extra attention. A question might be confusing because you do not know the subject matter to which it refers. If this is the case, don't eliminate the answer before you have affirmatively settled on another. When you come across an answer choice of this type, set it aside as you look at the remaining choices. If you can confidently assert that one of the other choices is correct, you can leave the confusing answer aside. Otherwise, you will need to take a moment to try to better understand the confusing answer choice. Rephrasing is one way to tease out the sense of a confusing answer choice.

7. Your First Instinct

Many people struggle with multiple-choice tests because they overthink the questions. If you have studied sufficiently for the test, you should be prepared to trust your first instinct once you have carefully and completely read the question and all of the answer choices. There is a great deal of research suggesting that the mind can come to the correct conclusion very quickly once it has obtained all of the relevant information. At times, it may seem to you as if your intuition is working faster even than your reasoning mind. This may in fact be true. The knowledge you obtain while studying may be retrieved from your subconscious before you have a chance to work out the associations that support it. Verify your instinct by working out the reasons that it should be trusted.

8. Key Words

Many test takers struggle with multiple-choice questions because they have poor reading comprehension skills. Quickly reading and understanding a multiple-choice question requires a mixture of skill and experience. To help with this, try jotting down a few key words and phrases on a piece of scrap paper. Doing this concentrates the process of reading and forces the mind to weigh the relative importance of the question's parts. In selecting words and phrases to write down, the test taker thinks about the question more deeply and carefully. This is especially true for multiple-choice questions that are preceded by a long prompt.

Quick Overview

9. Subtle Negatives

One of the oldest tricks in the multiple-choice test writer's book is to subtly reverse the meaning of a question with a word like *not* or *except*. If you are not paying attention to each word in the question, you can easily be led astray by this trick. For instance, a common question format is, "Which of the following is...?" Obviously, if the question instead is, "Which of the following is not...?," then the answer will be quite different. Even worse, the test makers are aware of the potential for this mistake and will include one answer choice that would be correct if the question were not negated or reversed. A test taker who misses the reversal will find what he or she believes to be a correct answer and will be so confident that he or she will fail to reread the question and discover the original error. The only way to avoid this is to practice a wide variety of multiple-choice questions and to pay close attention to each and every word.

10. Reading Every Answer Choice

It may seem obvious, but you should always read every one of the answer choices! Too many test takers fall into the habit of scanning the question and assuming that they understand the question because they recognize a few key words. From there, they pick the first answer choice that answers the question they believe they have read. Test takers who read all of the answer choices might discover that one of the latter answer choices is actually *more* correct. Moreover, reading all of the answer choices can remind you of facts related to the question that can help you arrive at the correct answer. Sometimes, a misstatement or incorrect detail in one of the latter answer choices will trigger your memory of the subject and will enable you to find the right answer. Failing to read all of the answer choices is like not reading all of the items on a restaurant menu: you might miss out on the perfect choice.

11. Spot the Hedges

One of the keys to success on multiple-choice tests is paying close attention to every word. This is never truer than with words like *almost*, *most*, *some*, and *sometimes*. These words are called "hedges" because they indicate that a

statement is not totally true or not true in every place and time. An absolute statement will contain no hedges, but in many subjects, the answers are not always straightforward or absolute. There are always exceptions to the rules in these subjects. For this reason, you should favor those multiple-choice questions that contain hedging language. The presence of qualifying words indicates that the author is taking special care with his or her words, which is certainly important when composing the right answer. After all, there are many ways to be wrong, but there is only one way to be right! For this reason, it is wise to avoid answers that are absolute when taking a multiple-choice test. An absolute answer is one that says things are either all one way or all another. They often include words like *every*, *always*, *best*, and *never*. If you are taking a multiple-choice test in a subject that doesn't lend itself to absolute answers, be on your guard if you see any of these words.

12. Long Answers

In many subject areas, the answers are not simple. As already mentioned, the right answer often requires hedges. Another common feature of the answers to a complex or subjective question are qualifying clauses, which are groups of words that subtly modify the meaning of the sentence. If the question or answer choice describes a rule to which there are exceptions or the subject matter is complicated, ambiguous, or confusing, the correct answer will require many words in order to be expressed clearly and accurately. In essence, you should not be deterred by answer choices that seem excessively long. Oftentimes, the author of the text will not be able to write the correct answer without offering some qualifications and

5

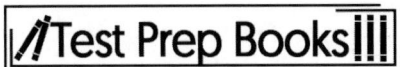

modifications. Your job is to read the answer choices thoroughly and completely and to select the one that most accurately and precisely answers the question.

13. Restating to Understand

Sometimes, a question on a multiple-choice test is difficult not because of what it asks but because of how it is written. If this is the case, restate the question or answer choice in different words. This process serves a couple of important purposes. First, it forces you to concentrate on the core of the question. In order to rephrase the question accurately, you have to understand it well. Rephrasing the question will concentrate your mind on the key words and ideas. Second, it will present the information to your mind in a fresh way. This process may trigger your memory and render some useful scrap of information picked up while studying.

14. True Statements

Sometimes an answer choice will be true in itself, but it does not answer the question. This is one of the main reasons why it is essential to read the question carefully and completely before proceeding to the answer choices. Too often, test takers skip ahead to the answer choices and look for true statements. Having found one of these, they are content to select it without reference to the question above. The savvy test taker will always read the entire question before turning to the answer choices. Then, having settled on a correct answer choice, he or she will refer to the original question and ensure that the selected answer is relevant. The mistake of choosing a correct-but-irrelevant answer choice is especially common on questions related to specific pieces of objective knowledge.

15. No Patterns

One of the more dangerous ideas that circulates about multiple-choice tests is that the correct answers tend to fall into patterns. These erroneous ideas range from a belief that B and C are the most common right answers, to the idea that an unprepared test-taker should answer "A-B-A-C-A-D-A-B-A." It cannot be emphasized enough that pattern-seeking of this type is exactly the WRONG way to approach a multiple-choice test. To begin with, it is highly unlikely that the test maker will plot the correct answers according to some predetermined pattern. The questions are scrambled and delivered in a random order. Furthermore, even if the test maker was following a pattern in the assignation of correct answers, there is no reason why the test taker would know which pattern he or she was using. Any attempt to discern a pattern in the answer choices is a waste of time and a distraction from the real work of taking the test. A test taker would be much better served by extra preparation before the test than by reliance on a pattern in the answers.

Audiobook Access

We host multiple bonus items online, including access to our audiobook. Scan the QR code or go to this link to access this content:

testprepbooks.com/bonus/servsafe

If you have any issues, please email support@testprepbooks.com.

Introduction

Function of the Test

This test ensures that candidates for entry-level food protection management roles are adequately prepared, both in terms of knowledge and other skill areas, to properly and safely perform their duties. ANSI has accredited the ServSafe exam against the Conference for Food Protection (CFP) Standards for Accreditation of Food Protection Manager Certification Programs. The test is recreated every five years using guidelines from the National Council for Measurement in Education, the American Psychological Association, and the American Educational Research Association.

Test Administration

The ServSafe Food Protection Manager exam can be taken online or in person, though both options must be proctored by a registered ServSafe Proctor. Online testing requires an acceptable hardware and software setup. Be sure to have a photo ID with signature with you at your examination location.

The exam is provided in several languages, though these are different between online and print versions of the test. The online test can be conducted in English, Spanish, and Chinese. The print version is available in English, Spanish, Korean, Chinese, French Canadian, and Japanese. A large-print version of the test is also available in print. Candidates who would like to request these or other accommodations should contact the test-givers ten days prior to the test.

Test Format

The ServSafe exam consists of 90 multiple-choice questions in seven subject areas, as described below. 80 of these questions are considered "operational" and apply to the candidate's score on the exam. The other 10 questions are considered "pilot" questions, which are used to design future tests but which do not impact the test-taker's score on the exam.

The allotted time to take the test is two hours, but the average length of time taken by candidates to finish the test is one and a half hours.

Content Area	Percentage of Questions (Number of Operational Questions)
Management of Food Safety Practices	10% (8 questions)
Hygiene and Health	15% (12 questions)
Safe Receipt, Storage, Transportation, and Disposal of Food	16.25% (13 questions)
Safe Preparation and Cooking of Food	18.75% (15 questions)
Safe Service and Display of Food	10% (8 questions)
Cleanliness and Sanitation	15% (12 questions)
Facilities and Equipment	15% (12 questions)

Introduction

Scoring

In order to pass the ServSafe Manager exam, the candidate must receive a score of 75% or higher. This percentage only counts correct and incorrect answers from the 80 "operational" questions on the exam. The 10 "pilot" questions are not counted and are only used for research. This 75% cutoff means that a candidate must correctly answer 60 out of 80 questions to successfully pass.

If the test is successfully passed, then the candidate will be granted ServSafe Food Protection Manager Certification. An eCertificate can be printed, or a certificate can be mailed to the candidate from ServSafe for a fee. The certification is valid for a five-year period, unless otherwise indicated by local regulations.

Online testing will provide immediate exam results, but there is some delay before results for in-person exams will be provided, normally around ten days. With either testing option, the overall score can be accessed online using your last name and either your score access code or exam session code.

If a test is not successfully passed, a candidate may retake the test. There is a limit of two attempts within 30 days; if this second attempt is unsuccessful, the candidate must wait an additional 60 days before taking another retest. Finally, a candidate may take the test a maximum of four times within any given 12 month period. The exam must be purchased again, separately, for each retest.

Study Prep Plan for the ServSafe Test

1 **Schedule** - Use one of our study schedules below or come up with one of your own.

2 **Relax** - Test anxiety can hurt even the best students. There are many ways to reduce stress. Find the one that works best for you.

3 **Execute** - Once you have a good plan in place, be sure to stick to it.

One Week Study Schedule

Day 1	Management of Food Safety Practices
Day 2	Hygiene and Health
Day 3	Safe Preparation and Cooking of Food
Day 4	Facilities and Equipment
Day 5	Practice Test #1
Day 6	Practice Test #2
Day 7	Take Your Exam!

Two Week Study Schedule

Day 1	Management of Food Safety Practices	Day 8	Safe Service and Display of Food
Day 2	Hygiene and Health	Day 9	Cleanliness and Sanitation
Day 3	Food Contamination	Day 10	Manual Washing
Day 4	Employee Illness	Day 11	Facilities and Equipment
Day 5	Safe Receipt, Storage, Transportation, and Disposal of Food	Day 12	Practice Test #1
Day 6	Food Storage	Day 13	Practice Test #2
Day 7	Safe Preparation and Cooking of Food	Day 14	Take Your Exam!

Build your own prep plan by visiting:

testprepbooks.com/prep

As you study for your test, we'd like to take the opportunity to remind you that you are capable of great things! With the right tools and dedication, you truly can do anything you set your mind to. The fact that you are holding this book right now shows how committed you are. In case no one has told you lately, you've got this! Our intention behind including this coloring page is to give you the chance to take some time to engage your creative side when you need a little brain-break from studying. As a company, we want to encourage people like you to achieve their dreams by providing good quality study materials for the tests and certifications that improve careers and change lives. As individuals, many of us have taken such tests in our careers, and we know how challenging this process can be. While we can't come alongside you and cheer you on personally, we can offer you the space to recall your purpose, reconnect with your passion, and refresh your brain through an artistic practice. We wish you every success, and happy studying!

Management of Food Safety Practices

Foodborne Illnesses

According to the Centers for Disease Control and Prevention (CDC), about one in six people become sick from foodborne illnesses each year in the United States, resulting in approximately 128,000 hospitalizations and about three thousand deaths yearly. Foodborne illnesses like *E. coli*, *Salmonella*, and botulism are caused by toxins or pathogens—disease-causing microorganisms—such as bacteria, viruses, parasites, and fungi. Each person involved in a product's journey from its source of origin—such as a farm—to the point where a consumer eats or drinks a product plays a role in ensuring that product is safe, including the consumers themselves. A **foodborne illness outbreak** occurs when two or more foodborne illness cases result from a common food source, and research shows that the majority of foodborne disease outbreaks have been linked to food service establishments. Food safety managers play a critical role in helping prevent foodborne illnesses, and this study guide explains key strategies that they can use to protect their customers.

Foodborne illnesses pose a greater risk for certain groups of people, called **highly susceptible populations (HSPs)**. These groups include preschool-aged children, people with a compromised immune system (such as from chemotherapy), elderly people, and people who eat food from certain institutions (such as day cares or nursing homes). Specific guidelines for these types of facilities help protect higher-risk groups from acquiring foodborne illnesses.

Agencies for Foodborne Illness Prevention

Because foodborne illnesses are such a prevalent national issue, there are a few government agencies that work to prevent them. Each agency has different roles in serving this purpose.

The USDA, or U.S. Department of Agriculture, inspects all eggs, meat, and poultry.

The FDA, or Food and Drug Administration, inspects all foods excluding eggs, meat, and poultry. It is also the issuer of the Food Code, which provides regulations that affect most restaurants and food service operations.

The CDC, or Centers for Disease Control and Prevention, works with food service operations to investigate ongoing foodborne illness outbreaks and offer education on foodborne illnesses and causes.

There are many other regulatory authorities that affect different jurisdictions, and they each have different food codes. This means that different operations will have to follow different rules regarding food. For example, hot-holding temperatures may differ from one state to another.

Time/Temperature Control for Safety (TCS) Foods

Most cooked foods and some raw foods can easily become vehicles that harbor and transmit pathogens and toxins to people, resulting in foodborne illness. Managing the temperature and amount of time these foods—referred to as **time/temperature control for safety (TCS) foods**—spend at a given temperature is necessary for limiting pathogen growth and toxin formation. Some examples of TCS foods are:

Management of Food Safety Practices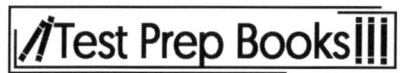

- Cooked or raw beef, pork, fish, shellfish, and chicken
- Eggs, milk, and dairy products
- Cooked fruits and vegetables
- Cooked grains, such as pasta and rice
- Raw, cut melons, tomatoes, sprouts, and leafy greens
- Ready-to-eat foods such as potato salad, pizza, and casseroles

In most cases, it is important to cook TCS food to the temperature recommended for that food and to limit the amount of time it spends in the **danger zone** between 41°F (5°C) and 135°F (57°C). Much like humans, pathogens do best when they are not too hot and not too cold, so they will thrive and multiply within the danger zone. The less time TCS foods are allowed to remain within this temperature range, the less likely they will be able to make someone sick. When cooling food below 135°F, the temperature must drop from 135°F to 70°F (21°C) within two hours and from 135°F to 41°F within six hours. If food is prepared with room-temperature ingredients, such as shelf-stable canned goods, the prepared food must be cooled to 41°F within four hours. Food with a greater mass or volume and food that is covered will take longer to cool, but cooling times can be reduced using a variety of methods, such as stirring the food in a container placed in an ice water bath, using rapid cooling equipment, or maximizing the exposed surface area by using larger or multiple containers. In circumstances where TCS foods are delivered to the food establishment, monitoring to ensure proper temperature control both upon arrival and during the transportation process also help ensure food safety. Ready-to-eat TCS foods that will not be sold or consumed within 24 hours must be clearly marked with the date by which they should be consumed or sold. When refrigerated, this is 7 days after being prepared.

Using time alone as a control factor, TCS foods can be held without temperature control if consumed or discarded within:

- 4 hours after being removed from refrigeration at or below 41°F or from cooking at or above 135°F, with a label or other indicator of the end of the 4-hour period

- 4 hours if the food is at or below 70°F when it becomes a TCS food—such as when a melon is cut or a canned good is opened—and remains at or below 70°F for the duration of the 4 hours, with a label or other indicator of the end of the 4-hour period

- 6 hours for refrigerated foods that start out at or below 41°F and are held at or below 70°F during the 6 hours, with a label or other indicator denoting both the beginning and the end of the 6-hour period

While it is important to be familiar with the guidelines above because they apply to a wide range of food, it is also important for food service employees to be familiar with exceptions and specific rules within their own facilities.

Thermometers

Thermometers are necessary to ensure foods are adequately cooked, as research has shown that the appearance of food is an unreliable indicator of its temperature or safety. Food thermometers must be accurate to within +/−1°C or, if there is no Celsius scale on the thermometer, to within +/−2°F and should be calibrated on a regular basis before each shift as well as after an impact or a drastic

temperature change, such as using with ice water or boiling water. There are two methods for calibrating a thermometer: the ice-point method and the boiling-point method.

The ice-point method is the safest and easiest way to calibrate a thermometer. It involves filling a container with ice (preferably crushed/shaven ice) and filling it the rest of the way with water. Then, a thermometer is put into the ice water; if it doesn't read 32°F, it should be calibrated.

The boiling-point method includes bringing a pot of water to a boil, then putting the thermometer in the water (without touching the pot). The thermometer should read 212°F, or it should be otherwise adjusted.

Correct thermometer use includes the following:

- Clean, sanitize, and air-dry thermometers before and after each use.
- Insert the entire sensing area—from ¼ inch to 3 inches, depending on the type of thermometer. The stem may need to be inserted sideways for thin foods like chicken strips, pork chops, and hamburger patties.
- Place the thermometer in the center or thickest part of most foods, such as casseroles, roasts, and individual cuts of meat or parts of poultry, avoiding bone and fat.
- Take temperature readings from multiple locations, particularly in whole poultry, foods that are not uniformly shaped, and combination dishes containing eggs, ground meat, or poultry.
- Avoid glass thermometers unless they have a shatterproof casing.

Thermometer	Usage
Thermocouple or thermistor thermometer	<ul><li>Quick, digital display</li><li>Sensor located in the tip of the probe</li><li>Can be used for foods of any thickness</li><li>Oven cord thermistor thermometers can be left in food during cooking</li></ul>
Bimetallic stemmed thermometer	<ul><li>Checks temperatures between 0°F and 220°F</li><li>Dial display</li><li>Insert up to the dimple, usually 2–3 inches</li><li>Good for larger and thicker foods</li></ul>
Infrared thermometer	Measures surface temperatures
Maximum registering thermometer	Displays the maximum temperature measured while being used
Time–temperature indicator (TTI)	Changes colors to indicate time–temperature abuse for stored food/deliveries

Risk Factors for Foodborne Illnesses

Data collected by the CDC consistently shows that five **foodborne illness risk factors** are the most significant contributors to foodborne illnesses from retail and food service establishments:

Management of Food Safety Practices

- Food from unsafe sources—such as those from unapproved facilities

- Inadequate cooking (not cooking food to a high enough temperature for a sufficient amount of time to kill or inactivate pathogens)—such as failing to cook fish to an internal temperature of at least 145°F for at least 15 seconds

- Improper holding times and temperatures—such as holding food in the danger zone for too long

- Contaminated equipment—such as using the same knife or cutting board for raw meat and then for cooked meat without proper cleaning and sanitation in between

- Poor personal hygiene—such as improper or infrequent hand washing

Based on data collected by the US Food and Drug Administration (FDA), food safety in food establishments is dependent on managers taking a proactive approach to controlling foodborne illness risk factors by incorporating specific procedures into the routine operations of the business and monitoring for correct use of those procedures, a concept referred to as **active managerial control**. Some ways to achieve active managerial control include:

- Using purchase specifications listing what items may be purchased and specific quality standards they must meet

- Conducting employee training programs, monitoring handwashing and glove safety

- Enforcing reporting policies for sick employees

- Implementing standard operating procedures (SOPs) and monitoring procedures

- Using recipes with specific instructions for ensuring that temperature and time requirements are met

- Monitoring the cooking, holding, and cooling times of TCS foods

- Managing and inspecting food deliveries

- Posting consumer advisories on the risks of raw or partially cooked foods

- Monitoring the cleaning and sanitizing procedures done by employees

The Hazard Analysis Critical Control Point (HACCP) system

The **Hazard Analysis Critical Control Point (HACCP) system** provides seven principles that can be used to achieve active managerial control of the five foodborne illness risk factors. The FDA endorses the use of the HACCP principles in general, but an HACCP plan is only mandatory for certain types of facilities. Most retail and food service establishments, however, can choose whether or not to use the HACCP principles as part of their food safety management system.

HACCP principles are meant to be implemented in addition to a strong foundation of **prerequisite programs**—a facility's standard procedures for protecting food from contamination, preventing overgrowth of bacteria, and maintaining equipment on an ongoing basis. These procedures may include

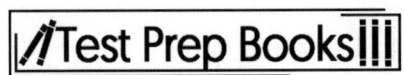

specific criteria or processes for vendor certification, employee training, and management of allergens; specific instructions within recipes; and other standard operating procedures (SOPs) such as First In, First Out (FIFO). With prerequisite programs like these in place to promote basic sanitation and safe operations, these seven HACCP principles further promote active managerial control of foodborne illness risk factors.

Principle 1: Conduct a hazard analysis.
The hazard analysis consists of two parts: hazard identification and hazard evaluation. Food safety hazards may include biological, chemical, or physical hazards.

The particular foods served and preparation processes involved at a given facility will determine what hazards could occur there. These hazards should then be evaluated to determine whether they should be included in the HACCP plan based on their likelihood of occurrence and the severity of the risk they pose. **Control measures** can then be identified to eliminate or reduce the risks of those hazards.

Principle 2: Determine the critical control points (CCPs).
A **critical control point (CCP)** is a step at which a control measure must be implemented to eliminate a hazard or to reduce it enough to prevent danger.

Principle 3: Establish critical limits.
Each critical control point has at least one **critical limit**, which is a number that delineates whether or not sufficient reduction or elimination of a hazard has been achieved by the control measure to ensure food safety, such as cooking ground beef to an internal temperature of 155°F for at least 17 seconds.

Principle 4: Establish monitoring procedures.
Monitoring procedures are measurements and observations that verify that critical limits are being met and food safety policies are being followed. They may include measuring temperatures or pH levels and observing whether employees are using gloves properly.

Principle 5: Establish corrective actions.
When a critical limit is not met, a corrective action may involve discarding the food or may provide a way to still ensure the food is safe—for example, by further heating the food until safe limits are met.

Principle 6: Establish verification procedures.
These are written procedures that lay out the details of how the efficacy of the HACCP plan will be verified, including the specific actions used in verification, who will perform those actions, and how often they will perform them. Verification procedures are distinct from monitoring and may include checking to ensure that monitoring procedures are performed properly, that corrective actions are taken when needed, that monitoring records are accurate and recorded at correct intervals, and that monitoring equipment is functioning properly—such as verifying thermometer calibration.

Principle 7: Establish record-keeping procedures.
Record-keeping procedures will establish how to document activities that fulfill prerequisite programs and HACCP plan procedures, and these records may prove vital in protecting the food establishment if it is investigated in a foodborne illness case.

Management of Food Safety Practices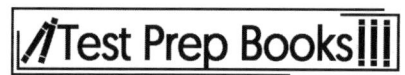

Responding to a Foodborne Illness Outbreak

There should be a crisis management team in place for every food service operation. This team may include one representative or multiple, depending on the size of the operation. Most likely, members include the owner, an employee from management (like a general manager, CEO, etc.), and the chef or food-preparing employee. Crisis teams may also include legal representatives, quality assurance agents, public relations officials, and more, especially for large organizations.

Preparing for a foodborne illness outbreak begins before food is even served. The crisis management team should have a plan set in place to prepare for any crisis events which may occur. The plan should contain emergency phone numbers and when to use them, including the number of every member of the crisis management team. It should also have phone numbers for the local fire department, police department, and other resources like testing labs and the jurisdiction's regulatory authority.

One person should be selected as the spokesperson for the food service operation to speak to the media. That way, the messages that go out are more consistent, rather than having multiple employees answer questions that they may not be prepared for. The media representative should have basic interview skills and an understanding of how to best respond to media inquiries.

The crisis response team should design a form to give customers who report a foodborne illness. The form should ask for the customer's contact information, descriptions on when and what the customer ate, and other pertinent data. Specific needs will vary based on each food service operation.

If only one customer reports a foodborne illness, their information should be gathered and the form that was created by the crisis response team should be filled out and returned. Workers at the food service operation, managers, and owners should never accept liability or responsibility for the customer's foodborne illness until an actual investigation is done and facts are presented.

If an outbreak is suspected, the following actions will help prevent additional cases and aid authorities in determining the cause.

- Collect important data from the ill persons, including symptoms, food eaten, and contact information using a pre-designed contact form.

- Close the facility to avoid additional exposure to whatever food may be causing infection.

- Contact the local regulatory entity, such as the city health department. Explain the issue, respond to follow-up questions fully and honestly, and encourage employees' full cooperation.

- Collect information from food employees and document who was working when the suspected outbreak occurred.

- Label, set aside, and document details about any food implicated in the illness, and avoid throwing away other food in case it is needed. All food that is suspected to be involved should be labeled with "Do Not Use" and "Do Not Discard".

- If any customers still have the contaminated food, ask for samples to use for testing.

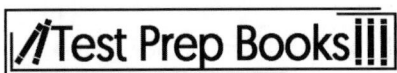

- If an employee is suspected as the cause for the foodborne illness outbreak, exclude them from working until they are cleared.

Responding to Other Crises

Aside from foodborne illnesses, other crises may affect a foodservice operation including water service interruptions, power outages, fires, floods, and other natural hazards. Any time there is a severe risk to the health and safety of employees or customers, operations must be stopped. However, food service operations are allowed to continue running in some circumstances as long as the operation has a written emergency operating plan in place that is approved by the local jurisdiction.

If the food service operation plans to continue operating during an imminent hazard, they must notify the regulatory authority that they will be implementing their emergency operating plan and take immediate action to prevent any food safety risks or health hazards.

Different hazards require different responses, and each should be noted in the emergency operating plan. One of the most common hazards is water service interruption. In the case of not having access to water, the following ideas may be considered as part of the emergency operating plan.

There should be a predesigned menu that includes items to be served that require little to no water. Bottled water should be stored and available to use in an emergency. Emergency contact information should be stored for local authorities and plumbers. There should be an emergency hand-washing plan in place. Have a supply of single-use items and condiments available. If restrooms are unavailable and there are no other restrooms available for staff to use, the operation must stop until restrooms become available.

In the case of a power outage, there are a few things that should be prepared ahead of time. First, there should be access to a generator and a refrigerated truck that can be used at any time in case of emergency. There should be a predesigned menu with foods that don't require cooking. Depending on what equipment is still able to run during a power outage, the operation may or may not have to close or stop cooking. If a power outage causes there to be no refrigeration, the time of the outage must be noted and the freezer and cooler doors should stay closed. If equipment used for hot holding isn't working, then the time of the outage must be noted. If the power outage lasts less than four hours, hot held foods may be reheated. If it lasts more than four hours, they must be thrown away. If ventilation hoods or fans do not work due to an outage, the operation must cease all cooking. Both during and after a power outage, it is crucial to temperature-check foods to ensure that they do not remain in the danger zone for longer than four hours.

Emergency events like fires require emergency response by the fire and police departments. Numbers for both local departments should be listed by the phone. Floods may require the response of plumbers, utility companies, and/or emergency services depending on the severity. In the case of a flood, it's crucial to keep people away from wet floors or anywhere they may slip and fall. If the flooded area affects the food or food preparation area in any way, all operations must be ceased. Any food or food packaging that got wet must be discarded.

Management of Food Safety Practices

Employee Training Best Practices

All employees must be trained in general food safety as soon as they are hired, and management should provide regular refreshers. This training includes handwashing, hygiene guidelines, preventing cross-contamination, cleaning, and other operation-specific tasks. Specific workers will also need additional food training.

Because each of the five major foodborne illness risk factors is a result of human error, food workers are the frontline defenders in protecting their customers from foodborne illness. While it would be great if people did everything they were taught to do, this is often not the case. Most people have heard of the importance of regular exercise, plenty of sleep, and eating their vegetables, yet ever-increasing rates of lifestyle-related diseases confirm what most know from experience: knowing doesn't always translate to doing.

Knowledge of food safety is only beneficial if the employee actually uses it. Research shows that giving employees information through knowledge-based training is not as effective in changing their behaviors as behavior-based training, which incorporates elements that motivate learners by addressing their attitudes and beliefs. In fact, during peak times, knowledge-based training alone can have no significant impact, while behavior-based training does.

Here are some strategies to improve the efficacy of food safety training.

- Tailor training to the audience by considering trainees' learning preferences, styles, and aptitudes. For example, incorporating a combination of auditory, visual, and tactile elements into the training helps engage students with different learning styles or with limited English proficiency.

- Explain why it is important to practice food safety activities and report illnesses, and consider sharing current, local foodborne illness cases.

- Avoid overly long training sessions.

- Schedule refresher training at least once per year.

- Incorporate ongoing observation and feedback to acknowledge compliant behaviors and correct noncompliance.

- Foster motivation and employees taking pride in their work through methods such as rewards and goal-setting.

- Provide mentoring to increase employees' confidence in their ability to put food safety measures into practice.

- Set a good example.

It is common for food handlers to believe that they must choose to either practice food safety or actually get their jobs done. This may help explain why food safety behaviors tend to decrease during peak times. Fostering a culture that emphasizes, reinforces, rewards, and models food safety practices is essential to preventing foodborne illness.

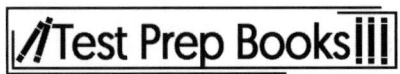

Inspections

Inspections are a regular occurrence in the food service industry. They ensure food is kept safe from contamination and that proper code is being followed. Inspections usually occur at least once every six months for most food service operations; however, some operations will need inspections more or less often.

Inspections will usually happen without any kind of warning, and operations should have a general plan for how to handle them. Inspectors may ask for documentation on a number of things, so it's important to have organized data on hand that covers cleaning schedules, food safety certificates, food information, and any logs kept.

There are three types of risk items that will be inspected: priority items, priority foundation items, and core items.

Priority items include the most important ways to prevent foodborne illnesses, like following correct procedures for glove usage, wound coverage, and handwashing.

Priority foundation items help make the priority items successful. Providing proper gloves for each employee, task, and use is considered a priority foundation item.

Core items are the main things to remember when it comes to cleaning and sanitizing, like keeping areas clean and equipment sanitized.

After the inspection, the inspector will give a final score and explain any violations or corrections that must be made. Violations must be corrected within 72 hours for priority items and within 10 days for priority foundation items. In rare cases, health inspectors can issue an immediate suspension of all operations due to extreme health violations. If this happens, the operation must pass a reinspection in order to reopen after fixing all issues.

To avoid having issues arise during inspections, it's important that managers do self-inspections of their food service operations routinely. They should check for the same issues that health inspectors look for and fix any problems that are noted.

Hygiene and Health

Foodborne Illness Symptoms and Signs

Foodborne illnesses can cause an array of symptoms that can occur within 30 minutes or up to six weeks after consuming contaminated food. Although each person may experience difference symptoms depending on the cause of their foodborne illness, most people report at least one of the following common symptoms:

- Vomiting
- Nausea
- Diarrhea
- Fever
- Abdominal cramps

Food Contaminants

There are three types of hazards that can contaminate food: biological, chemical, and physical.

Hazard	Examples
Biological hazards	<ul><li>Bacteria, viruses, and parasites</li><li>Toxins produced by these pathogens</li></ul>
Chemical hazards	<ul><li>Naturally occurring toxins, such as food allergens, mycotoxins, fish and shellfish toxins, and mushroom toxins</li><li>Chemicals from products like cleaning solutions, pesticides, and beauty products</li><li>Chemicals added to process or preserve food</li><li>Metals that leach into food from kitchen utensils and vessels</li></ul>
Physical hazards	<ul><li>Fragments of glass, jewelry, bone, or fingernails</li><li>Metal from cans or staples</li><li>Other small objects that could cause choking, cuts, or infection</li></ul>

Pathogens that Cause Foodborne Illnesses

The pathogens that cause foodborne illnesses are a diverse group categorized by type of organism: bacterium, virus, parasite, or fungus.

Bacteria

Many types of bacteria can cause foodborne illnesses. While most bacteria thrive between 41°F and 135°F, some can grow in refrigeration temperatures, some form spores that can survive cooking as part of their life cycle, and some produce toxins that are also heat resistant. The amount of time it takes from ingesting a contaminated food to developing symptoms—the **incubation period**—is most commonly around 12–72 hours, but it can range from 30 minutes to several weeks. Common symptoms include abdominal cramps, nausea, vomiting, fever, and diarrhea that is often watery and, for certain illnesses, bloody. The most important precautions in preventing illnesses caused by bacteria are purchasing food only from approved sources, avoiding cross-contamination of ready-to-eat (RTE) food with juices from raw meat and poultry, cooking food to safe temperatures, washing hands, and excluding sick food

Hygiene and Health

employees from work. The following bacteria, often referred to as the **Big Six**, are especially risky in a food establishment setting.

Bacteria	Example associated foods
Norovirus	• Salads • Fresh fruit
Shiga toxin-producing *E. coli* (STEC)	• Ground beef • Produce
Hepatitis A	• Berries • Shellfish • Salads
Salmonella Typhi	• RTE foods • Drinks
Nontyphoidal *Salmonella*	• Animal products • Produce
Shigella	• Salads such as potato salad • Produce

Viruses

The main viruses responsible for foodborne illnesses are noroviruses and hepatitis A, which are commonly associated with RTE foods and spread primarily via the fecal–oral route; this means virus particles making their way from one person's feces to another person's mouth, either directly from person to person or indirectly through contaminated food or drink. Infected employees, even if asymptomatic, can shed viral particles through their feces that may remain airborne in the restroom for hours, and this is one reason proper hand washing, avoiding bare-hand contact with RTE food, and excluding sick employees from work are such critical steps for preventing foodborne illnesses.

Additionally, bivalve molluscan shellfish—such as clams and oysters—may harbor norovirus and hepatitis A due to contamination from feces in the water they grow in and should only be purchased from approved suppliers. Some viruses, including hepatitis A, can survive cooking temperatures.

Norovirus is the number one cause of foodborne illness in the United States, causing approximately nine hundred deaths per year and leading to the hospitalization of one in 160 children before the age of five. The virus can survive on surfaces for up to two weeks even when subjected to freezing, many chemical disinfectants, and heating to insufficient cooking temperatures. Its symptoms include vomiting, diarrhea, abdominal cramps, nausea, fever, and body aches.

Parasites

Parasites are organisms that get their nutrition from a host such as a human or other animal. They spread in the form of a cyst and are usually associated with wild game, fish, or products that have been contaminated through water. To prevent parasitic illnesses, food should only be obtained from approved sources and should be cooked or frozen properly, depending on the specific food and use.

Toxoplasmosis, an infection with the parasitic protozoan *Toxoplasma gondii*, is the fourth leading cause of hospitalization and second leading cause of death due to foodborne illness in the United States. It is especially risky for women who are or may become pregnant and for individuals with compromised

Hygiene and Health

immune systems. Other notable parasitic illnesses in the US include cryptosporidiosis, cyclosporiasis, giardiasis, and trichinellosis.

Fungi

Fungi is a classification that includes molds, mushrooms, and yeasts. Molds are microscopic fungi that exist virtually everywhere, grow on plants and animals, and produce spores that travel through the air to plants and animals, where they become established and continue to grow. Most species are harmless, but some molds produce toxins—called **mycotoxins**—that can cause illnesses in humans and animals. Food that is visibly moldy should be thrown out.

A certain amount of mold contamination from the natural environment is considered unavoidable, but government regulatory agencies monitor food products to ensure levels are not too high. One example of a mycotoxin that is monitored by the FDA is aflatoxin, produced by some types of *Aspergillus* molds, which grow on crops such as nuts and grains (like corn and rice). Milk may also contain aflatoxin if the cow ingested it while eating grains. In severe cases, too much aflatoxin can lead to cancer, liver failure, or death.

Mushrooms should only be purchased from reputable suppliers to ensure that the right species is picked. Most cases of food poisoning involving mushrooms occur when they are purchased from unreliable sources that mistake toxic mushrooms for edible ones.

Conditions Leading to Bacterial Growth

Bacteria grow best in six conditions, often referred to by the acronym FAT TOM. This stands for food, acidity, temperature, time, oxygen, and moisture.

Food is the first key condition in bacterial growth, because bacteria need the nutrients within food to survive. TCS foods, which are listed in the previous section, are especially vulnerable to bacterial growth.

Acidity is the next key condition for bacterial growth because bacteria grow better on less acidic foods. That means that foods like tomatoes, vinegar, and lemons have a lower chance of harboring bacteria, while bread, milk, and raw meat have a higher chance of bacterial growth.

There is a temperature danger zone for certain foods (from 41°F to 135°F) which allows for the growth of bacteria. Foods that should avoid being kept in the temperature danger zone are time-temperature control for safety (TCS). Some foods that are TCS include poultry, dairy products, sprouts, and sprout seeds.

Time is another key condition when it comes to bacteria growth. Since it takes time for bacteria to grow, the longer the food sits within these conditions, the more bacteria will grow on it.

While some bacteria need oxygen to grow, others rely on the absence of oxygen for their growth.

Moisture is the final component in bacterial growth, with bacteria thriving in foods with high moisture content.

Bacterial Growth Stages

Once all six conditions for bacterial growth are met, bacteria will grow in four stages: lag, log, stationary, and death. Lag is the first stage of bacterial growth, when the bacteria is first introduced to the food. At this phase, the number of bacteria remains about the same as it waits for ideal growth conditions. The next stage is called the log stage, at which point bacterial growth is supported and bacteria can double as quickly as every 20 minutes. After this growth stage, bacteria hit a stationary stage where their numbers will stay about the same. At this time, the conditions will have changed to limit bacterial growth, which will eventually lead to the final stage, the death stage. At this time, the bacteria population declines as bacteria die more quickly than they can grow.

Fish and Shellfish Toxins

Fish and shellfish can accumulate biological toxins, or poisons, from their environments and food, so their toxicity usually depends on their source of origin and feeding habits. Fish and shellfish toxins cannot be destroyed or inactivated by freezing or cooking, so certain types of fish from certain locations—including locations with temporary safety advisories—should be avoided by receiving fish only from safe sources that provide proper labeling. Unlike most fish toxins, scombrotoxin—a toxic level of histamine—develops after harvesting and can be prevented by maintaining fish under refrigeration beginning as soon as possible after catching or harvesting the fish.

Symptoms of fish and shellfish poisoning usually begin within a few minutes or a few hours of ingesting the toxin and resolve within a few hours or days, but some toxins can lead to long-term symptoms or death. Some symptoms are listed below.

- Nausea, vomiting, diarrhea, and abdominal pain
- Burning, numbness, or tingling sensations in or around the mouth, throat, or extremities
- Headache, dizziness, or a feeling of floating
- Joint or muscle pain, fatigue, or weakness
- Swelling, itching, rash, or dry throat and skin
- Difficulty walking, thinking, seeing, speaking, swallowing, or breathing
- Irregular heartbeat or low blood pressure
- Sweating; facial or upper body flushing
- Hot–cold inversion (hot food seems cold and vice versa)
- Paralysis, convulsions, seizures, or coma
- Tooth pain or metallic taste
- Pupil dilation

Hygiene and Health

Toxin	Common Sources	Additional Details
Histamine (Scromboid Poisoning)	Tuna, mahi-mahi, mackerel, bluefish, sardines, anchovies, marlin, bonito, skipjack, herring, marlin, amberjack	• Fish sometimes have an unusual appearance (honeycombed), smell, or taste (peppery, metallic, sharp, or salty).
Ciguatoxin	Barracuda, moray eel, grouper, snapper, sturgeon, sea bass, amberjack, mackerel, parrot fish, triggerfish, hogfish	• It can cause death. • Barracuda and moray eel should be avoided.
Tetrodotoxin	Pufferfish (also known as fugu or blowfish)	• It can cause death. • These fish should be avoided.
Saxitoxin, Brevetoxin, Domoic Acid (Shellfish Poisoning)	Mussels, clams, oysters, scallops, whelks, and certain gastropods	• Paralytic shellfish poisoning (caused by saxitoxin) is the most likely to occur and the most high-risk; death can occur in as little as 30 minutes if enough of the toxin is consumed. • Neurotoxic shellfish poisoning (caused by brevetoxin) is often associated with red tides and is usually not life-threatening. • Amnesic shellfish poisoning (caused by domoic acid) is rare, can cause short-term memory loss, and very rarely can cause death.

Chemical and Physical Contaminants

Foodborne illnesses are mainly caused by either a foodborne infection or a foodborne intoxication. **Foodborne infections** are illnesses in which biological pathogens themselves—like *E. coli* bacteria or the hepatitis A virus—make a person sick.

Foodborne intoxications are illnesses caused by chemicals and other toxins, including:

- Naturally occurring toxins that accumulate in seafood from the water and from other animals it ingests

- Toxins produced by pathogens, including *Bacillus cereus, Clostridium botulinum, Staphylococcus aureus*, and *Aspergillus*

- Chemicals added to foods unintentionally through the use of chemicals at any point in the production process, including pesticides, sanitizers, and cleaning products

- Substances added to food as preservatives

- Heavy metals from the environment—like mercury—or those like copper and lead that leach into food from utensils or from the vessels in which it is stored

The symptoms of foodborne intoxications often involve vomiting and diarrhea and typically develop more quickly than symptoms of foodborne infections.

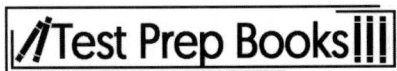

In addition to foodborne infections and foodborne intoxications, consumers may also be in danger if physical contaminants—including bone fragments and foreign objects—are introduced into the food within the food establishment or in some previous supply chain step (e.g., in its source environment, during processing, during packaging). This creates risks like broken teeth, choking, cuts, bleeding, infection, and the need for surgical removal.

Food Contamination

Contamination of food with bacteria, viruses, parasites, or biological toxins (biological contamination), chemicals (chemical contamination), or other hazards—such as the physical hazards (physical contamination) mentioned previously—occurs when these contaminants are transferred to the food from its surroundings, from another food, or from a person, which may include employees or consumers. Here are some steps to avoid contamination.

- Receive food only from sources that obtain food from uncontaminated sources and that ensure protection from contamination during processing and transport—such as those on an approved source list or approved brands.

- Inspect food upon delivery to verify that it is delivered at an appropriate temperature (usually at or below 41°F or at or above 135°F) and does not appear to have been subjected to unsafe temperatures at any point before delivery.

- Store food in a dry, clean place, at least 6 inches off the floor, away from any chemicals. It should also be a location where it is protected from splashes (such as from water during hand washing or from chemicals), from dust and other debris, and from overhead drips (such as condensation from an air vent).

- Keep food enclosed in its original packaging or within another container or wrap.

- Separate foods that can cross-contaminate one another. For example, separate raw animal products from ready-to-eat (RTE) foods and unwashed fruits and vegetables. Also, separate different types of animal products from one another unless they are ingredients in the same recipe.

- Use different equipment and utensils—such as knives, cutting boards, and containers—for foods that may cross-contaminate one another. Alternatively, clean and sanitize these items between uses on different foods.

- Practice proper hand washing, glove use, and personal hygiene.

- Exclude sick employees from work when appropriate.

- Prevent contamination by consumers by, for example, providing individually packaged condiments, not re-serving uneaten bread or chips, and ensuring clean utensils are used for buffets.

Intentional Food Contamination

Along with accidental or unintentional food contamination, food can also be intentionally tampered with. Whether it's a disgruntled employee, the competition, or even a terrorist, there are many people who may try to intentionally contaminate food.

Managers can use the acronym ALERT to create a food defense program for their food service operation. It stands for assure, look, employees, reports, and threat.

Assure means that managers should be sure that the products they purchase are from approved, safe suppliers. Managers should supervise all deliveries and make sure all food storage areas (including delivery vehicles) stay locked at all times.

Look means that managers should keep their eyes on products in their food service operation. This may include limiting access to food preparation and certain areas, monitoring the proper storage of food and chemicals, and keeping an eye out for anything out of the ordinary.

Employees means that managers should be diligent about knowing their employees, conducting background checks, and making sure that only authorized employees are present in certain areas.

Reports mean that managers should keep data involving their food defense program, including files on staff, food/products received, and other documents. The data should be added to regularly and accessible when needed.

Threat is the final component of a successful food defense program. This means that managers should have a plan in place to deal with any possible threats to their food service operation.

Food Allergens

There are nine **major food allergens:** milk, eggs, fish, crustacean shellfish, tree nuts, wheat, peanuts, soybeans, and sesame. The majority of food allergy reactions are caused by an immune system response to proteins in these foods or foods derived from them. Symptoms usually appear within anywhere from a few minutes to two hours and may include:

- Itching, rashes, or hives
- Dwelling of some part of the body, such as the lips, tongue, or face
- Breathing difficulty, cough, or throat tightness
- Feeling dizzy or lightheaded
- Gastrointestinal distress, such as vomiting, nausea, abdominal pain, or diarrhea
- Anaphylactic shock, which can result in death

Even a very small amount of an allergen can be problematic or dangerous, and cross-contact of allergen-containing foods with other foods is a common cause of reported food allergy reactions. Studies indicate that while employees may be very happy to accommodate allergy-related requests, many do not have adequate training about food allergies, and some believe that a very small amount of the allergen will not cause a reaction. Having a system in place to clearly and accurately inform customers of the presence of potential food allergens—by providing them with ingredient lists or recipes, for example—and to prevent cross-contact—by preparing their meals in a freshly, thoroughly cleaned or separate, allergen-free area, for example—will enable them to safely enjoy their meal.

Hygiene and Health

Hand Washing, Glove Use, and Personal Hygiene

Hand Washing:

Proper hand washing, a critical practice for preventing foodborne illness, involves washing the hands, exposed arms, and any prosthetic devices used for these areas at the right times and in the right way, as determined by current research. The following steps should be taken to wash these areas for a total of at least 20 seconds, after which a hand antiseptic that meets FDA guidelines can be applied and allowed to air-dry as an optional step.

- Use soap and warm, running water to wash in a dedicated hand washing sink (not one used for food preparation, washing food equipment, or other purposes, such as cleaning mops).

- Rinse, apply soap, and then vigorously rub these areas for 10 to 15 seconds, making sure to remove debris underneath fingernails and to include the fingertips and areas between fingers.

- Rinse off soap.

- Dry completely.

- Use a barrier such as a dry paper towel to turn off the faucet and open the bathroom door.

Always wash hands immediately before handling food or clean equipment that will be used with food, before putting on gloves to work with food, any time hands are soiled or could cause cross-contamination, and after the following:

- Visiting the restroom
- Handling raw food, if switching to handling ready-to-eat food
- Coughing, sneezing, or using a tissue
- Eating, drinking, or using tobacco products
- Touching body parts other than clean hands or arms, such as the face
- Touching animals, such as a personal service animal or aquatic animals used in a display
- Touching dirty equipment or utensils
- Using cleaning chemicals
- Leaving and returning to the food preparation area
- Handling anything that could cause contamination, such as money, trash, or cell phones

Glove Use:

To prevent contamination, food employees should never handle ready-to-eat (RTE) food with bare hands except:

- When the food is added to a dish that will be cooked to at least 145°F or minimum cooking temperature (for raw meats)

- When the jurisdiction or regulatory authority allows bare hand contact with RTE food

Instead, single-use gloves or another utensil such as tongs, a serving spoon, or deli tissue should be used. To ensure that single-use gloves are effective for preventing contamination, it is important to:

- Wash hands first
- Choose the right size.

Hygiene and Health

- Put them on properly (touching only the edge, not blowing into them or rolling them up)
- Change them when they become dirty or damaged, when switching tasks or types of food (such as after handling raw animal products, before handling ready-to-eat foods, or after an interruption to food preparation activities), or at least every four hours.

Personal Hygiene:

Food employees should also adhere to these personal hygiene practices, which are important to prevent contamination of food that can lead to foodborne illnesses.

- Wear clean clothes and bathe regularly.
- Wear hair restraints, such as hair nets, hats, and beard restraints.
- Avoid wearing jewelry on the hands or arms, except a plain ring.
- Keep fingernails smooth. Avoid fingernail polish and artificial nails unless covered by gloves.
- Securely cover any wounds or sores with discharge. If the lesion is on an exposed arm or hand, cover with an impermeable barrier and additionally—if on the hand or wrist—with a single-use glove.
- Avoid eating, drinking, and using tobacco products in food preparation areas.
- Food employees who have discharge from the eyes, mouth, or nose due to ongoing coughing, sneezing, or runny nose should be restricted as explained below.

Employee Illness

To reduce the risk of spreading pathogens, employees must report certain symptoms and diagnoses—referred to as **reportable symptoms** and **reportable diagnoses**—to managers, who then determine whether to **exclude** the employee from the food establishment for a period of time or to **restrict** the employee from working directly with food and food contact equipment such as utensils. When a food employee has jaundice (yellow eyes or skin) or is given a reportable diagnosis—an illness caused by norovirus, hepatitis A, *Shigella*, Shiga toxin-producing *E. coli* (STEC), *Salmonella* Typhi (typhoid fever), or nontyphoidal *Salmonella*—with or without symptoms, the regulatory authority must be notified and must give approval before the employee returns to work.

Additionally, food employees who have potentially been exposed to the following pathogens—through household contact or involvement with a confirmed disease outbreak—within a certain amount of time must be restricted from facilities primarily serving a highly susceptible population (HSP).

- Norovirus—within the past 48 hours
- STEC or *Shigella*—within the past 3 days
- *Salmonella* Typhi and Nontyphoidal *Salmonella*—within the past 14 days
- Hepatitis A—within the past 30 days

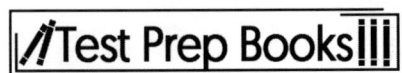

Hygiene and Health

Reportable Symptom	Actions
Vomiting or diarrhea	• Exclude until asymptomatic for at least 24 hours OR cleared by medical documentation
Jaundice	• Exclude if jaundice first appeared within the past 7 days unless cleared by medical documentation
Sore throat with fever	• Exclude from facilities primarily serving an HSP • Restrict in facilities not primarily serving an HSP • Return to work with medical documentation
Lesion with pus	• Restrict unless properly covered

Safe Receipt, Storage, Transportation, and Disposal of Food

Using Approved, Reputable Suppliers

There are many people involved in the food supply chain, from farmers and growers to manufacturing, packing, trucking, and distribution companies. Managers must ensure that their suppliers meet all federal, state, and local laws and regulations. Suppliers must undergo and pass all required safety inspections and should be able to provide evidence of these inspections. Managers can also see the suppliers' inspection reports by contacting the US Food and Drug Administration (FDA) and the US Department of Agriculture (USDA). The inspection report should include information regarding the suppliers' receiving, storage, and processing practices; cleanliness; staff training; and compliance with HACCP, GMP, GAP, or other safety guidelines.

Working with well-known, respected suppliers is the first step to ensuring that the food is free of defects and disease and is safe for human consumption. Developing an ongoing relationship with suppliers and learning about their food safety practices is a solid best practice. One of the most common safety practice guidelines is the Hazard Analysis Critical Control Points (HACCP). HACCP is a food safety management system that is recognized and used worldwide. It assesses food safety through analyzing and controlling physical, biological, and chemical risks. HACCP applies to all stages of food production, starting with raw materials all the way through to handling, manufacturing, and distribution. HACCP guidelines are based upon the foundations established by the Good Manufacturing Practices (GMPs; also referred to as the current Good Manufacturing Practices, or cGMPs).

Food safety practices should be based on the GMPs set forth by the Food and Drug Administration, which are designed to ensure good quality and safety of food products. GMPs are part of the US Code of Federal Regulations (21 CFR 111) and include the five Ps: people, procedures, products, premises, and processes.

The people involved in the process should be well-trained, have clear responsibilities, and be routinely assessed to ensure they are meeting all safety and quality standards. Manufacturing procedures should be standardized, transparent, and routinely reviewed and updated as needed. The products, including both the raw ingredients and the final product, should meet all quality standards and be closely monitored throughout the manufacturing process. The premises, including all equipment used, should be regularly inspected and kept clean and in good working order. Finally, the processes used throughout manufacturing should be continuously reviewed and updated.

The USDA sets forth Good Agricultural Practices (GAP) guidelines for growing produce. GAP guidelines are voluntary, but most reputable farmers and growers opt to adhere to these practices, as they are designed to prevent foodborne illnesses by reducing microbial contamination on farms. The GAP guidelines can be found in the FDA's publication *Guidance for Industry: Guide to Minimize Microbial Food Safety Hazards for Fresh Fruits and Vegetables*.

Documents and Tagging Guidelines for Food

Certain foods require documents with their delivery. For example, shellfish must be delivered with a shellstock identification tag. The tag includes information like the harvest date and location, as well as the shipping date. Tags must be held on file for 90 days after the last shellfish is used.

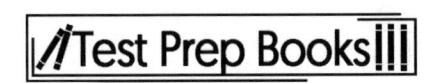

Like shellfish, any fish that will be used for sushi or other raw dishes must be delivered with the proper identifying documents. The documents include information about how the fish was frozen and stored, and, if the fish were farm-raised, additional documents are required stating that the fish meet FDA standards. Any documents that come with the sale of the fish must be kept for 90 days afterward.

Meat and poultry should also be delivered with the proper documentation, which includes an inspection stamp from the USDA and an identification number.

Planning and Receiving Deliveries and Supplies

It is important to inspect deliveries as soon as they arrive. Do not schedule deliveries for times when the manager or staff will be away, busy, or otherwise unavailable to inspect the delivery upon receipt. Employees responsible for delivery receipt and inspection should be properly trained and provided with the necessary equipment to inspect the delivery. This includes any purchase orders or receipts needed to verify that the correct products and amounts are being delivered, scales for weighing produce and other products, and thermometers to ensure items are being kept and delivered at the appropriate temperatures.

Receivers should look at the delivery truck to check for any indications of contamination. Packaging should be inspected to ensure that it is clean, undamaged, and that there are no signs of mistreatment. Use-by dates should also be verified and current. When receiving a delivery that includes time or temperature control for safety (TCS) foods, checking the temperatures of these items is imperative to ensure that they have traveled and arrived at the proper food-safe temperatures.

Receiving Temperature Requirements

Cold food items such as meats should be 41°F or cooler. Check fresh meat and poultry by placing a thermometer directly into the thickest part of the meat. Measure vacuum-sealed by placing the thermometer between two packages or, if possible, by folding the package around the thermometer, taking care not to puncture the packaging. Measure other items, such as yogurt, by inserting the thermometer directly into the product without touching the packaging.

Live and shucked shellfish and milk should arrive at an air temperature of 45°F or cooler and at an internal temperature no greater than 50°F; the temperature must be lowered to 41 degrees or cooler within four hours. Shell eggs should be received at a temperature of 45 degrees or cooler. Frozen foods should be frozen solid upon receipt, and hot foods should have temperatures of 135°F or warmer. Milk should be received at an internal temperature of 45°F or cooler, and should be chilled to 41°F or colder within four hours. Frozen foods that have water or fluid stains on the packaging or have ice crystals on the food should be rejected, as this is evidence of possible thawing and refreezing. When the shipment has been received and approved, the food items, particularly refrigerated and frozen items, should be stored in the appropriate areas immediately.

Key Drop Deliveries
Sometimes deliveries must be made before or after working hours, when managers and staff may not be on the premises to receive the delivery. In these circumstances, a supplier is given a key or access to the premises to make the delivery. The supplier places the products in freezers, refrigerators, or dry storage areas. This is known as a key drop delivery and should only be permitted from well-respected, trusted suppliers. Key drop deliveries should be inspected immediately when the manager or staff arrives. The receiver should check that the products were delivered to the appropriate storage areas, that they are

free of contamination and were protected from contamination when placed in storage, and that they appear to have been delivered with care and according to the purchase order and delivery agreement.

Rejections and Recalls

Sometimes an item or items from a delivery must be rejected. Products that have damaged or torn packaging or packaging that shows evidence of contamination or tampering should be rejected. This includes cans that have bulges, dents, rust, swollen ends, or illegible and/or missing labels, as well as sealed packages that are bloated or leaking. Items with broken seals, damaged cartons, or packaging that is dirty or discolored should also be rejected. Products that show signs of leakage, dampness, water stains, or damage should be rejected, as should any product that shows signs of pests. Additionally, any product that is beyond the use-by or expiration date and any product missing its freshness date should be rejected. Inspect food quality and reject any foods that show signs of mold, pests, discoloration, inappropriate texture (such as sliminess or stickiness), or odors.

Managers and receiving staff must know the proper procedures for handling rejected items. The item should be placed away from the rest of the delivery immediately. Report the issue to the delivery driver and include a detailed description of exactly what is wrong with the item. The delivery driver should provide a signed receipt of adjustment or credit before the item is returned to them. The receiver should also log the incident into the delivery record, either directly on the invoice or purchase order or in the store's delivery records according to the procedure set forth by the manager.

Occasionally, food items may be recalled by the manufacturer. This occurs when there is a problem with the manufacturing process, with the raw ingredients, or with the shipping and handling of the goods. Recalls may also be issued if products are incorrectly labeled, such as omitting allergen warnings that should be present on the packaging. The FDA and USDA issue recall notices, but vendors and suppliers should also notify their customers of any recalls.

Recall items can be identified using the information provided in the recall notice, such as product identification numbers and date, time, and/or location stamps. Once a recall item has been identified, it should be immediately removed from inventory and secured away from the usual food storage areas to prevent accidental use. The item should also be kept away from food prep areas, including utensils and equipment. The item should be clearly marked to avoid accidental use, such as placing a "do not use" sign on the product. The vendor will have specific instructions on how to handle the item, such as disposing of it immediately or returning the item to the vendor.

Specific Guidelines for Rejecting Foods

There are different guidelines for what should be accepted and rejected when inspecting specific foods.

Meat (including beef, lamb, and pork) should have no odor and firm flesh that isn't slimy. Beef should be bright red, although aged and vacuum-packed beef may be a darker red or even purple. Lamb should be light red, and pork should be light pink with white fat. Brown colors on any cut of meat is a concern and a reason to reject it.

Poultry should be the same light pink/white color uniformly, with no discoloration in any areas. It should have firm flesh with no stickiness or odors.

Shell eggs should be received clean, unbroken, with no odor. Otherwise, they should be rejected.

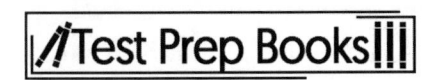

Dairy products should have no signs of mold and be a uniform color. They should have a mildly sweet taste and firm texture. Fresh produce should show no signs of mold or mishandling to be accepted, though color and flavor may vary.

Fresh fish should have bright red gills and firm, shiny skin, clear and bright eyes, and a mild ocean smell. Fish with an excessively fishy odor, cloudy eyes, soft flesh, or dull gills should be rejected.

Shellfish should be unbroken, closed, and alive (if received fresh). A fishy smell, slimy or dry texture, or broken shells are all reasons to reject shellfish. Crustaceans should also be received alive. Crustaceans that are dead or have an excessively fishy smell should be rejected.

Food Storage

In terms of storage, it is important to properly handle food from the moment that a shipment is received. Once a shipment of food is received, the foods should be carefully examined. The temperatures of TCS foods and frozen foods are particularly important to check. If TCS foods are determined to be safe, they should be quickly stored. Frozen foods should be frozen when they arrive. One sign that a food has thawed is that ice crystals have formed on the food's packaging.

In addition to ensuring that food deliveries are in good condition, managers must also ensure that all the food and products in their facilities are stored correctly to maintain quality and safety. One of the most important aspects of food storage is proper labeling. Food must be labeled to prevent confusion (such as mistaking one food item for another) and allergen contamination. Items in their original containers are likely already sufficiently labeled, but items that are removed from their original containers must be properly labeled. Labels should include the common name of the item, such as "flour," as opposed to a brand name that might not be immediately recognized. Labels could also include a photo of the item for easy identification. While it may not always be necessary to label foods that could not be mistaken for anything else, labeling everything is a good habit for safety.

In addition to labeling for identification, labels should also include freshness date information. Even refrigerated foods can contain bacteria, such as *Listeria monocytogenes*, that is harmful and can cause illness. Ready-to-eat TCS foods must be labeled carefully and must include date information if the food will be held longer than 24 hours. Ready-to-eat TCS foods cannot be stored for longer than seven days and must maintain a temperature of 41°F or cooler. The seven-day time frame begins on the day the food was prepared for consumption or the date the sealed container was first opened. The "prepared," "use by," or "discard" date should be clearly indicated on the label. Foods that contain ingredients with varying use-by dates should be marked according to the earliest date. For example, if a container of potato salad is prepared on March 20, but the mayonnaise used in the salad has a use-by date of March 25, then the use-by date on the label should be March 25, even though that is earlier than seven days from the date of preparation. It is always better to err on the side of caution regarding freshness dates.

Just as foods should be received at the proper temperatures, foods should also be stored in appropriate areas and at appropriate temperatures to maintain freshness. Dry goods should be stored in areas free of moisture, pests, and other contaminates. Goods should be stored at least 6 inches off the floor and away from walls. Containers should be sealable, airtight, and should protect the items from fluids. Chemicals, such as cleaning products, must always be stored in appropriate, airtight containers, and should never be stored near food items. Additionally, food should never be stored in non-food-specific areas such as locker rooms, mechanical rooms, under stairwells, or in restrooms or garbage collection areas.

Safe Receipt, Storage, Transportation, and Disposal of Food

In storing food from a shipment, cross-contamination should be avoided. Ensure that foods are kept separate and held properly. One practice that helps with the storing of food is keeping older food more accessible to ensure they are used before going bad. Also, raw foods should not be stored on top of foods that are ready-to-eat to avoid raw foods dripping down onto the ready-to-eat foods. Foods should be held outside of the temperature danger zone. Make sure that hot-held and cold-held foods stay within the appropriate safe temperature ranges. To do so, the temperatures of these foods should be monitored.

Storage Temperature Requirements

TCS foods should be stored at a temperature of 41°F or cooler. Meat, fresh fish, shellfish, and poultry should all be stored at an internal temperature of 41°F or cooler. Eggs should be stored at an air temperature of 45°F or cooler. Whole fruits and vegetables should be stored between 60°F and 70°F. Food that is vacuum packed (or reduced oxygen packaged (ROP) food) should be stored at 41°F or lower. Frozen foods should be stored at a temperature low enough to keep them frozen solid. Raw meats and seafood should be wrapped or covered and stored away from produce and ready-to-eat foods to avoid cross-contamination. If separate containers or coolers are not available, raw produce and ready-to-eat foods can be stored on higher shelves in the cooler to prevent any drips from the meats from contaminating them. Refrigerated foods should be stored in a particular order if they all share the same cooling unit. Produce and ready-to-eat foods should be on the top shelves. Beneath those should be seafoods, followed by whole cuts of pork and/or beef. Ground meats and fish should be below that, and ground poultry should be on the bottom shelves. Meat should be stored as far from the door as possible to ensure that it stays cold.

Certain best practices should be in place to ensure that coolers and freezers are in optimum working condition. There should be at least one thermometer or thermostat in each unit to confirm adequate temperatures are being maintained. Those thermometers or air temperature measuring devices must be accurate to within 3°F or 1.5°C.

Coolers and freezers should not be overloaded, as this can restrict airflow and cause the unit to have to work harder and be less efficient at maintaining an adequate temperature. Similarly, shelves should be open wire shelves free of any type of liner or covering to allow for proper air flow. Food items should be randomly spot-checked to ensure that sufficient temperatures are being maintained.

Stock Rotation and Food Disposal

Food items should be rotated to maintain quality and freshness, as well as to prevent spoilage and waste. Foods with the earliest use-by or discard by date should be used first, before their expiration dates. Foods with later expiration dates can be stored behind foods with upcoming expiration dates so the new foods can be rotated in as older supplies are used. This is referred to as the first-in, first-out (FIFO) method. Foods that are received first are used first, and foods received later are used after earlier stock is consumed. Once the expiration date arrives, unused foods should be disposed of, and newer stock should be moved forward.

Safe Preparation and Cooking of Food

When it comes to food preparation, the recommended guidelines for handling various kinds of food are important. First, it is important to know more general guidelines for food preparation. Personal hygiene is very important in food prep to maintain safe and sanitary practices to help avoid foodborne illness. Hands should be washed with warm soapy water for at least twenty seconds to ensure their cleanliness before and after handling food. There should be reminders on display for employees regarding handwashing. Hair should be kept up and away using hair restraints or hats. Jewelry on the hands should generally be avoided, and uniforms should be clean. Employees' food and drink should be kept away from both food and the food preparation area.

Staff should be trained to properly clean and sanitize surfaces, equipment, and utensils. The first step to clean and sanitize surfaces is removing or scraping off any food that is on the surface. Before sanitizing the surface, it should be cleaned. This should be done with a cleaning solution and can aid in removing leftover residue. After a cleaning solution is used, the surface should be rinsed in order to remove the solution and ensure the surface can be properly sanitized. The next step is to sanitize the surface. Different restaurants use different sanitizing solutions (such as sanitizers that are quaternary-based). The sanitizer should be spread with a disposable wipe, not a paper towel or reusable cloth; they can both absorb the sanitizer and limit how well it can be spread across the surface. The sanitizer should be spread over the surface and allowed to air dry for at least one minute.

To clean and sanitize equipment, first shut off and unplug the equipment. Any removable parts can be removed and then washed, either with the dishwasher or by hand. Just like when cleaning and sanitizing surfaces, equipment surfaces must be wiped of debris before being washed with a cleanser. This cleanser should be rinsed off for the sanitizer to be applied to the equipment. The equipment and its removable parts should then be allowed to air dry before putting them back together.

Fruits and vegetables should be rinsed before handling. Raw foods, such as meat, should not be rinsed as this can cause bacteria to spread through water onto other surfaces. Raw foods should be kept separately from other foods to prevent cross-contamination. This applies to food being stored and cut since raw foods should be cut on a separate cutting board. Also, marinades should not be reused unless they were first brought to a boil. Food must be stored and cooked properly. Food should be cooked to its minimum internal temperature for safety.

Although commonly forgotten, ice is also a food and must be kept sanitary and safe. Some ways to do that include using drinkable water to make ice, storing ice scoops outside of the ice machine, and never touching ice with bare hands.

Cooling and Reheating Foods

One area of concern with holding food is cooling. The best way to properly cool foods is through two steps. Foods should reach 70°F within two hours and, following this, should reach 41°F within four hours (total time from the beginning of the cooling process).

If the food has not cooled to 70°F within two hours, it should instead be reheated to minimum cooking temperatures and then cooled again. The total cooling time to reach 41°F should not exceed 6 hours total, or the remaining food must be disposed of.

Requirements for reheating food depend on what will be done with the food afterwards. If it's to be immediately served, the food can be heated to any temperature as long as it was first properly cooked and cooled. If the food is to be hot-held and then served, it must be heated to at least 165°F for 15 seconds first.

Thawing Foods

Foods that are being marinated should be placed in the refrigerator. Food should also be thawed properly, not just taken out and placed on the counter. There are three different ways to properly thaw frozen food: the refrigerator, the microwave, or cold water. Thawing food in the refrigerator is a more time-consuming method. The time required to thaw food fully and safely is important to consider when choosing which method to use. Larger items take more time to thaw, sometimes up to several days. The thawing time can also be affected by factors such as what temperature the fridge is set to and varied temperatures at different areas within the same fridge. After food is thawed, it is safe to stay in the fridge for additional time, depending on what the food is. Food can also be thawed in the microwave. When thawing food in the microwave, some parts of the food may begin to cook (but not fully), which makes it dangerous to hold and store. Because of this, when using the microwave to thaw food, it should be cooked immediately following thawing. The third safe method of thawing food is by submerging the food in cold, running water, though this should not be done in the three-compartment sink used to wash dishes. It must be sealed so that the bag it is in cannot leak, thereby preventing the food from collecting more bacteria. Water must be at 70°F or lower and should be running at a steady flow. This process takes at least an hour, depending on the weight of the food. Once the food is thawed using this method, it also must be cooked immediately.

Some foods can also be thawed during the cooking process, like frozen chicken (when deep fried) or frozen hamburger patties. Depending on the recipe, the food may be used directly from the freezer or it may be slacked (allowed to thaw slightly) before using. Either way, these foods should always be temperature checked to ensure that they reach minimum cooking temperatures.

Eggs and Egg Mixtures

In preparing eggs and egg mixtures, surfaces should be cleansed both before and after coming into contact with raw eggs and egg mixtures. To ensure that eggs are fully cooked, the whites and the yolks must be firm. Some foods use eggs that are raw or not fully cooked. In this situation, eggs and egg products that have been treated via methods such as pasteurization must be used to prevent the contraction of *Salmonella*.

Most batters and breading contain eggs or milk, making them more susceptible to time and temperature abuse. When using a batter or breading, be sure to store any extra at 41°F and place food that has been breaded back into the fridge immediately. It's important to remember to temperature check food that has been breaded because the thickness of the batter may make the cooking times longer than expected.

Rinsing Produce

It is important to inspect your produce before preparation and consumption. Any produce that appears rotten should be thrown out, and damaged parts of the produce should be removed. When preparing produce, it should be thoroughly rinsed. Produce does not need to be washed with soap in order to be cleaned, and certain chemicals are approved as produce rinses. Even if the skin of the fruit is not going to be eaten, it should still be rinsed. If the fruit has a rind, such as cantaloupe, it should be scrubbed. This is done to prevent bacteria from transferring to the inside of the fruit upon being cut or peeled. After being rinsed, produce should be dried to further lessen the bacteria on its surface. This can be done with a paper towel or a clean towel. It's important to make sure that produce never comes in contact with surfaces that have been exposed to raw meat.

Proper Cooking Temperatures

It is important to know the minimum internal temperatures that various foods must be cooked to for safe consumption.

For any kind of raw chicken and turkey, the minimum internal temperature is 165°F. Pasta, stuffed meat, and stuffing should also reach 165°F.

Ground pork, beef, lamb, and veal must maintain a minimum internal temperature is 155°F for 17 seconds to be considered safe. Mechanically tenderized meat like brined ham and shell eggs that will be hot held for service must also reach 155°F for 17 seconds.

For roasts of pork, beef, lamb, and veal, the minimum internal temperature is 145°F, which must be maintained for four minutes.

All types of fish and shellfish, as well as shell eggs that will be served immediately, must reach a minimum internal temperature of 145°F for 15 seconds. Steaks and chops of pork, beef, lamb, and veal must also reach 145°F for 15 seconds.

Plant-based foods like vegetables and grains should be cooked to 135°F. Tea should be served at 175°F.

Partial Cooking

Sometimes, food is cooked partially, stored, and then cooked again just before being served. In this case, the food must not be cooked for longer than an hour, then cooled immediately after. On the second cooking, the food must meet minimum internal cooking temperatures.

Special Requirements When Prepping and Cooking Food

Certain food service operations will have special requirements, which are issued through a variance by their regulatory authority. A variance notes that a certain requirement is changed for a specific food service operation. The following food preparation methods require the issuance of a variance:

- Packaging food using reduced oxygen packaging, or packaging fresh juice to sell at a later time
- Using additives (like vinegar) to preserve food
- Preserving food through curing or smoking
- Sprouting seeds
- Processing animals for personal use (for hunters, etc.)

Safe Service and Display of Food

Safe Service of Glassware and Dishes

There are various types of glassware and dishes, and it is important to know how to properly handle them. Generally, glassware and dishes should be handled gently. One important thing to remember about glass is to be careful about temperature. Glass cannot move too quickly between very hot and very cold temperatures since this can lead to glass breaking (thermal shock). Glass must be allowed to reach room temperature before being brought to very hot or very cold temperatures. Glass can also be made more susceptible to damage through coming into contact with other objects. It is generally impossible to tell the integrity of glass by visual means; it is important to attempt to limit glass coming into contact with other objects to maintain its integrity. Glassware should not be carried in multiples at a time to avoid damage through contact. Stacking glasses should also be avoided since this could result in many glasses being broken if the pile of glasses is knocked over; in addition, stacking glasses causes the glasses to come into contact with one another. There are also things to keep in mind regarding washing glassware. Glassware should be allowed to cool after coming out of the dishwasher to help prevent thermal shock from occurring. Another important thing to keep in mind for washing glassware in a dishwasher is to ensure that glassware is placed on the correct rack and to keep glassware from touching each other. Cold glasses should be allowed to warm up to room temperature by dumping out any ice remaining in them before going through a dishwasher or washing by hand. When washing dishes by hand, avoid letting glassware rest in the water. This prevents the glass from hitting the bottom of the sink or other dishes and glassware within the sink. Glasses should not be used if they are cracked or have a chip in them. Glasses should also be stored properly. Dishes, glasses, and utensils must be handled with care, particularly when they are being served. Plates should be held in the palm, and fingers should be placed under the plate or along the edge of the plate, avoiding contact with food. Utensils should be held only by the handle. Glasses should never be carried by their rim.

In addition to risks like breaking and cracking, dishes and glassware are also at risk of contamination by employees. It's important to remember that dishes should be carried by the edges or bottom to avoid contaminating food-contact areas. Glasses and flatware should only be held by the handles and employees should avoid touching any part that comes into contact with food. When serving dishes, if it's necessary to move or touch food that is ready to eat, the employee must use utensils or wear gloves (preferably both) to avoid bare hand contact.

Holding Food

Facilities that serve ready-to-go hot food—such as buffets—may hold hot food for up to four hours. During this time the food must be kept at a minimum of 135°F, although temperatures of 141°F and higher are preferable to keep the food out of the danger zone. Appropriate devices for maintaining heat include steam tables, holding cabinets, or heated trays. Food should be inspected and stirred frequently. Inspection improves safety by maintaining employee awareness of when the hot food has expired. Stirring helps the food remain consistently at a safe temperature. When hot food is not monitored, the manager risks serving unsafe food due to their or their employees' negligence.

Food intended to be served cold—such as a salad—must be stored at a consistent temperature of 41°F or below. After removing the cold food from the refrigerator, the food is considered safe for up to six hours as long as its temperature does not rise above 70°F. Food service workers should utilize cold tables, cold crocks, and similar devices to ensure ready-to-go cold food remains as close to 41°F as possible. However, if the food is only held for four hours, it can reach any temperature as long as it is discarded at the four-hour mark.

The temperature of both hot and cold food should be checked by an employee at least every two hours. Any prepared cold meal or cold tray which has been out for too long or reached too high a temperature must be discarded.

It's important to implement time management practices in the facility. Maintaining a consistent schedule of when food is set out and when employees must check food reduces the risk of causing foodborne illness. Different practices are best for different facilities. It is the manager's job to establish temperature checks as part of the food safety regimen and ensure that they are carried out by the employees.

Food Re-Service Guidelines

Sometimes customers return meals because they aren't what was ordered or expect. Other times, a table may leave an entire container of uneaten bread or chips and salsa. Food that is served to a guest that is returned to the kitchen in any circumstance should NOT be re-served. Even if the food appears to be uneaten or is simply a garnish or condiment, it should never be re-served to other guests or combined with other foods to be re-served.

The only foods that may be re-served includes condiments in their original packaging (ketchup/mustard bottles), and unopened, prepackaged foods (like single-serve condiments or crackers).

Customer-Facing Service

Many measures must be taken to maintain the cleanliness and safety of self-service bars. Any self-serve areas should be monitored closely to ensure cleanliness. When food is brought out to the self-serve area from the kitchen, it should be covered as it is brought out. When replacing food, new food should not be mixed with old food; rather, the whole container should be replaced. Certain things should not be placed within the container, such as food that has been previously handled, objects, and the handles of serving utensils. When it comes to serving utensils, only clean utensils should be used in serving food. Anything that is contaminated, whether it be utensils or food, should be removed immediately. Food should be protected from coughs and sneezes to help prevent it from becoming contaminated. One way to protect food from this is to have a barrier such as a sneeze guard. Customers should be encouraged to use sanitary practices through means such as signs posted around the self-serve area to make these practices clear. Other sanitary practices for customers to participate in involve dishware and utensils. Plates should not be used more than once. Utensils that will be used for eating should be grabbed by the handle, and they should be stored in a way that customers first grab them by the handle. Finally, temperatures of foods should be monitored to ensure they are staying within safe zones for consumption. Cold foods should be held below 41°F. Hot foods should be held at 135°F and above. For each shift, there should be a food service employee who is assigned to supervise and watch for contamination in self-service areas.

Open containers and condiments should be monitored for their cleanliness. When empty, open bottles of condiments should be replaced, not refilled. Some condiments, due to contamination risks when they are served in open containers, must be thrown away at the end of the business day. Certain condiments must be refrigerated during non-business hours. The shelf lives of opened and unopened condiments must be monitored as well. Mayonnaise can last for around three months unopened at room temperature and, once opened, must be refrigerated and used within two months. Mayonnaise cannot be left opened and unrefrigerated for more than two hours; this runs the risk of contamination. Certain condiments are safe to keep unrefrigerated once opened, but they do not last nearly as long as refrigerated condiments. For example, ketchup can be unopened at room temperature for around one year. Once opened, it can last for around six months refrigerated. However, if ketchup is opened and unrefrigerated, it can last for around one month.

Off-Site and Catering Service Guidelines

Often, a food service operation will participate in off-site events like catering, delivery, or pop-up stands. Each of these different food service methods will require employees to prepare the food onsite and then take it offsite while keeping it safe. When it comes to catering, it's important to remember to use insulated containers, ice, and other methods to control the temperature of the food. Pop-up stands, or temporary food service units, are usually set up for events and will usually only operate in one location for less than two weeks. It's important that safe drinking water is available for any off-site events for both handwashing and cooking purposes.

If the food service operation also operates a vending machine, it's important to check the foods in the machine daily to make sure they are still within the use-by date.

Cleanliness and Sanitation

While the terms *cleaning* and *sanitizing* are sometimes used interchangeably, it is important to understand the difference between them. Cleaning removes surface dirt, debris, food particles, and other undesirable material from surfaces. Sanitizing involves the use of chemicals to remove bacteria from surfaces. Disinfecting goes a step further and removes bacteria and viruses from surfaces. Specific products are needed for sanitization and disinfecting surfaces, while usually a combination of water and soap or detergent is sufficient for cleaning. Cleaning is usually enough for non-food-contact surfaces, such as walls, floors, and storage shelves. Sanitization is required for any surface, utensil, or equipment used in food preparation. All food contact surfaces should be cleaned and sanitized when changing to prepare a different kind of food, whenever a task is interrupted and the surface may have become contaminated, after use, and every four hours during continuous use.

There should be a master cleaning schedule that includes what needs to be cleaned, how often, and who is responsible for cleaning it. It should also include detailed instructions on how to do each cleaning task.

Cleaning

To clean surfaces, any stuck-on materials need to be removed using a brush, pad, or clean cloth. Then the surface should be washed using an appropriately mixed cleaning soap or detergent. The surface should be scrubbed with some vigor, making sure that the soap fully covers the surface. Next, rinse the surface with clean water. Depending on the surface, this can be done by pouring clean water over the surface or by wiping the surface with a clean, wet cloth. Once the surface is rinsed, it is usually a good idea to dry the surface with another clean cloth.

There are four major types of cleaners: detergents, delimers, degreasers, and abrasive cleaners. Each type of cleaner serves a different purpose, but may share some ingredients with each other. Detergents can be used for most cleaning tasks; they work to remove dirt and grime from surfaces. Delimers are specifically made to remove mineral deposits and other types of residue, often left by hard water and buildup. Degreasers have components that help to dissolve grease, and are usually used in areas that get very greasy. Abrasive cleaners are usually used to scrub off food or stuck-on objects from other equipment.

There are two types of wiping cloths: wet and dry. Wet cloths are used with sanitizing solution to wipe down surfaces and equipment. Dry cloths are used for food spills. It's important that the two cloths are only used for their specific purposes.

Sanitizing

Heat sanitization and chemical sanitization are the two primary methods of sanitizing surfaces. Heat sanitization involves heating the surface to at least 171°F and soaking the surface for at least thirty seconds. This type of sanitization works best for items that can be submerged, such as utensils. Chemical sanitizers can be applied by soaking the surfaces or by spraying, swabbing, or otherwise applying the chemical to the surface. The most common sanitizing chemicals are chlorine, iodine, and quaternary ammonium compounds. Chemical sanitizers are closely regulated by the EPA and should always be used carefully according to the manufacturer's directions.

Cleanliness and Sanitation

Sanitizers must be used correctly to be effective. Key aspects of correct use include the concentration of the sanitizer, temperature, pH, water hardness, and the length of time the sanitizer is in contact with the surface.

Concentration

Concentration involves the appropriate mix of sanitizer and water. Concentration is measured in parts per million (ppm) and should be checked against the test kit that is usually included with the product. Having the correct concentration is critical to the effectiveness of the sanitizer. Too much water results in a weak mixture that may not effectively sanitize the surface, and too little water results in a strong mixture than could be hazardous. Always mix sanitizer in accordance with the manufacturer's instructions.

Temperature

Sanitizing concentrate should be mixed with water that is a suitable temperature to ensure that the chemicals are sufficiently diluted. Using water that is too cold can result in a separated solution, while using water that is too hot could cause the chemicals to break down and become less effective. Water temperatures should be between 68°F and 100°F, depending on the chemical being used. If using iodine, the proper water temperature is 68°F; for quats, the water temperature should be 75°F. Mixing chlorine, the water temperature should be at least 75°F for water that has a pH of 8 or less, and at least 100°F for water that has a pH of 10 or less.

Water Hardness and pH

Water hardness refers to the level of minerals in the local water, and pH refers to the level of acidity in the water. This information can be obtained from the local water company, and the sanitizer manufacturer can provide information as to the suitable concentrate for these conditions.

Time

The final key to effective sanitization is making sure that the sanitizer is in contact with the surface for enough time. The amount of time necessary to fully sanitize a surface depends on the chemical being used. For chlorine, it's at least seven seconds, and for both iodine and quats it's at least thirty seconds.

To sanitize a surface, the surface must first be cleaned using the techniques described above. Once the surface is clean, a properly mixed sanitizer should be applied. The manufacturer's label should include directions for mixing the concentration into a usable solution as well as instructions for applying the sanitizer. It is important to make sure that the sanitizer thoroughly covers the entire surface. Once the surface has been fully sanitized, allow it to air dry.

Cleaning and Sanitizing Equipment

The correct way to clean and sanitize equipment depends on the type of equipment. Thoroughly review the manufacturer's instructions and recommendations prior to cleaning the equipment to ensure that nothing is damaged by incompatible cleaning products and that no one is injured through misuse of the equipment.

Any motorized or electrical equipment must be unplugged for safety prior to cleaning. Remove any parts of the equipment that can be removed for cleaning, such as the blades on a slicer. Wash, rinse, and sanitize those by hand or by running them through a dishwasher, if permitted.

For immovable parts of the equipment, scrape any food scraps away. Wash the equipment using a brush, towel, or other appropriate cleaning tool. Rinse the surface using another clean cloth or towel. Apply the sanitizing solution according to the manufacturer's instructions and ensure that the solution comes into contact with all the equipment's surfaces. Allow to air dry before reassembling the machine.

Machines that hold and dispense ready-to-eat TCS foods, such as ice cream machines and soda machines, should be cleaned daily according to the manufacturer's directions and local regulations. This usually involves running a cleaning and/or sanitizing solution through the machine.

Dishwashing

Dishes, flatware, and glass can be run through the dishwasher. Both a chemical and high-temperature dishwasher sanitize dishes, although through different methods. For items that contact food but are unable to be washed in the dishwasher (such as pans), a three-compartment sink can be used. The first compartment is used for cleaning, the second for rinsing, and the third for sanitizing. First, any food that is left on the dishes should be scraped off. Then, the dishes can be cleaned by scrubbing them in warm and soapy water. After this, they can be rinsed in the second compartment with warm water before being placed in the sanitizing compartment. The water in the first and second compartments must be a minimum of 110°F. After they are sanitized, they should be air dried.

Dishwashers clean and sanitize tools and equipment using both heat and chemicals. High-heat dishwashers heat the water to at least 165°F for stationary-rack machines and at least 180°F for all others. The dishwasher should have a built-in thermometer that shows the high temperature, and care should be taken to ensure that the machine is in correct working order. Otherwise, it may not effectively sanitize the dishes. Usually, plates, utensils, and small equipment parts can be cleaned in a dishwasher, while large pots and pans and other types of equipment must be cleaned and sanitized by hand.

As with other types of equipment, dishwashers must also be kept clean and well maintained. The machine should be checked daily to ensure that it is free of food and debris, that all spray nozzles are clean and in good working order, and that there are no mineral deposits.

Dishes must be properly prepped and loaded into the dishwasher for optimal cleaning. Scrape away any remaining food debris, ensuring that there are no stuck-on foods. Load the dishes so that the spray will reach all the dish surfaces, and never overload the machine. When the wash and rinse cycles have finished, allow the dishes and utensils to air dry. Drying with a towel or other cloth could cause recontamination.

Manual Washing

Tools and equipment, such as large pots and pans and serving platters, may require handwashing. This should be done using a three-basin sink. Each basin should be thoroughly cleaned prior to washing dishes. The first basin should be filled with an appropriate mixture of soap and water at least 110°F. The second basin should be filled with clean water or left open if a sprayer will be used to rinse the dishes. The third sink basin should be filled with the appropriate sanitizing mixture or suitably hot water for sanitizing. It is also helpful to have a clock with a second hand nearby so sanitization times can be monitored. Once dishes are washed, rinsed, and sanitized, they should be allowed to air dry to avoid recontamination.

Cleanliness and Sanitation

Pest Control and Management

Pests are a major health hazard in any food service facility. There are three key steps to avoiding pest problems: preventing access to the facility; depriving the pests of food, water, and shelter; and working with a reputable, licensed pest control company in the event of a problem.

The key concept to pest control currently favored in the food service industry is **integrated pest management**. This perspective encourages management to view pest reduction and elimination as an ongoing process rather than as an occasional task. Managing pest control is primarily about sanitation. While this may seem paradoxical at first, it makes sense considering the behavioral reasons pests infest a facility. If potential sources of food, water, and shelter are eliminated, then it's extremely unlikely that the facility will need to hire an exterminator or utilize hazardous pesticides. This both reduces expenses and improves safety.

The management aspect of integrated pest management is an important task. It is the manager's duty to create a system which is appropriate for their particular facility, with the result that there are fewer pests—hopefully none! They should delegate weekly, monthly, and quarterly sanitation checks and pest-reduction practices, and utilize documentation to ensure pest management is being completed. The manager should perform their own checks on occasion to confirm that the documentation is accurate.

The best prevention, of course, is keeping pests out of the building in the first place. Carefully check all shipments for evidence of bugs or rodents. Check the truck for rodents and/or bugs, alive or dead, as well as droppings. Check the packaging for evidence of pests, such as chewed or torn packages, egg cases, droppings, or body parts. Any shipment that shows evidence of pests should be immediately refused.

Pests can also enter the building through cracks, vents, and other unscreened openings. Check the building thoroughly for any accessible opening. Vents and windows should be covered with wire mesh screens that are small enough to prevent entry from even the tiniest bugs, at least 16 mesh per square inch. Vents should not be covered with any solid material such as metal caps, as those can prevent necessary air flow. Cracks in walls or floors should be sealed, as should any gaps or openings around pipes, plumbing, electrical, and HVAC lines. Air curtains or fly fans should also be used in external doorways to prevent bugs from entering when the doors are open. Self-closing devices on doors can also make it more difficult for pests to gain access to the facility.

Pests can also be deterred by a lack of water, food, and shelter or nesting space within a building. All food should be stored in airtight containers and on shelves at least six inches off the floor or in coolers or freezers. Spills should be cleaned up immediately, and garbage should be taken outside and stored in appropriate trash containers. Garbage containers should be kept clean and tightly covered. Cleaning the entire facility regularly will eliminate food sources for pests, which will make the facility a less-desirable home for them.

Types of Pests

Most common pests in food service facilities can be loosely thought of as either rodents or insects. All pests are drawn to what kitchens and other facilities provide: food, water, and shelter.

Mice and rats are common and persistent rodents. Mice are the smaller of the two, typically three or four inches in length. In contrast, a rat can easily grow to ten inches in length and weigh a pound or more. Both pests are characterized by their teeth and their droppings. All rodents' teeth constantly grow, requiring the animal to gnaw to keep the length manageable.

Consequently, signs of chewing, tearing, or scratching in wood and plastic can indicate the presence of mice or rats. Small black droppings are also solid evidence of a rodent infestation. Mice droppings are small enough to be mistaken for dust or debris, requiring additional scrutiny. Meanwhile rat droppings are a bit larger and less likely to be overlooked. Another important difference is that mice get most of their water through food, but rats typically require a source of standing water.

A great number of pests can plague a food service facility. The most common are:

- Cockroaches: These reddish insects are usually one to two inches long, and are found in gaps or cracks in the building and in warm, moist places. They typically carry pathogens like *Salmonella* Typhi, viruses, parasites, and fungi.

- Flies: Easy to overlook because they're so common, flies are just as able to carry illness as other insects. They're drawn to standing water and decaying material found around sources of trash or sewage. They are known to carry pathogens like *Shigella* spp.

- Ants: Small insects, typically black or red, which are drawn toward sources of food. Some bite, but typically the biggest danger is contamination of foodstuffs.

- Rodents: Rats and mice can carry many diseases and do considerable damage to food and property. They reproduce often and can reach infestation numbers quickly.

Different pests are drawn to different sources of food and water, but all find it easy to meet their needs in kitchens and other facilities. Maintaining a high standard of sanitation proactively reduces the likelihood of pests entering the facility by removing their sources of food and water.

Even the best facilities can still sometimes experience pest problems. The smallest amount of evidence can indicate a much larger pest problem. Where there is one, there are often many. When this happens, the manager should contact a licensed, reputable pest control operator (PCO) immediately. A pest control company can help eliminate the pests, identify the source of the infestation, and provide additional suggestions and products to aid in prevention of future problems. Treatment for pest control problems will vary depending on each food service operation's needs.

Pesticide Label Law

The use of chemicals to prevent or exterminate pests is governed by the **Federal Insecticide, Fungicide, and Rodenticide Act** (FIFRA). This act requires food service facilities to only use pesticides which have been registered as acceptable by the Environmental Protection Agency (EPA). Under FIFRA, all registered pesticides must have a label which explains their intended purpose, how to use the pesticide appropriately, potential hazards caused by the chemical, and safety precautions as well as basic first aid. There should also be a Safety Data Sheet (SDS) on hand that has corresponding information on any pesticides kept.

It's important that all employees and any contractors hired by the facility use pesticide chemicals with caution. Even when following label directions, it is possible for sprayed or dusted chemicals to contaminate food. Ensure employee safety through the use of protective equipment as described on the

Cleanliness and Sanitation

label, and ensure food safety by airtight sealing of any food stored in a space where chemicals will be used. If possible, temporarily move food to an alternative storage location. After chemicals have been used, thoroughly sanitize all food storage spaces and food preparation spaces to ensure the chemical does not contaminate anything which will be served to a consumer.

Always remember to check for additional regulations concerning the use of chemicals in the workplace and in food preparation within your state, city, or other local jurisdiction.

Facilities and Equipment

Poor maintenance, broken and/or damaged equipment, and a building that is in disrepair can all lead to serious problems with food and employee safety. These conditions can also result in health code violations, which, if serious enough, can shut down a facility. It is imperative that the building, facilities, and all equipment be kept in good working order and clean and safe condition to provide healthy food service.

If new construction is in order, it's important to first check with the regulatory authority for approval of the construction plans. There are certain requirements for how construction should be done and which contractors can be used.

Facilities and Equipment

Flooring, walls, and ceilings should be smooth and durable for better cleaning and maintenance. Porous surfaces harbor bacteria and are much more difficult to keep clean and sanitary. Walls, floors, ceilings, and window and door surrounds should be free of cracks, gaps, holes, or other openings where pests may enter. Cracked or broken floor or ceiling tiles should be replaced immediately, both to keep pests out and to avoid a safety hazard. Most food service operations use **coving** in their flooring, which is a curved tile between the floor and the wall rather than the typical sharp corners. The coving makes it easier to clean the tile and to keep dirt and grime from becoming trapped.

Some great, resilient flooring options include rubber tile, vinyl tile, quarry tile, marble, and wood. While carpeting may work for certain areas, it's not ideal for places that get very dirty or have heavy traffic.

Floor-mounted equipment, such as shelving, dishwashers, coolers, freezers, sinks, and counter units should be installed at least 6 inches, or 15 centimeters, off the floor to allow for proper cleaning and maintenance underneath. When the floors are cleaned and sanitized, the areas underneath this equipment should be cleaned and sanitized as well. Inspect floors regularly for damage, including inspecting underneath the equipment. Alternatively, floor-mounted equipment can be sealed to the floor to prevent contamination underneath.

Tabletop or countertop equipment must be on legs that are at least 4 inches, or 10 centimeters, high to allow for proper cleaning underneath the equipment. As with floor-mounted equipment, tabletop equipment can also be permanently affixed and sealed to the countertop.

All equipment that comes into contact with food must meet the appropriate national standards as established by NSF, which is accredited by the American National Standards Institute (ANSI). Under the NSF guidelines, food service equipment must be made of nonabsorbent, corrosion resistant, smooth material. It must also be durable, easy to clean, and damage resistant. For this reason, food preparation equipment is often made of stainless steel, which meets all these criteria. Cutting boards can be made from wooden or synthetic material, but it must be nonabsorbent and nontoxic.

All equipment should be regularly inspected and maintained by qualified technicians. Establishing a regular maintenance schedule, both in-house and with a technician, can help to ensure that the equipment is always in top working order. In addition, any equipment that appears to be malfunctioning and/or sustains damage should be immediately checked by a qualified technician.

Facilities and Equipment

Cleaning stations are critical to maintaining healthy, hygienic facilities. Dishwashers should be installed in a convenient, easily reachable location and should be installed according to the manufacturer's specifications. Dishwashers should be cleaned and maintained regularly. Always use detergents that are approved by the local regulatory agencies. Ensure that the dishwasher uses suitably hot water, 165 to 180°F depending on the machine, and has sufficient water pressure to thoroughly clean and sanitize the dishes and utensils. Suitable dishwashers should come equipped with the ability to measure water pressure and temperature (in increments no greater than two degrees) as well as the concentration of cleaning and sanitizing chemical agents. Most hot water heaters cannot heat water to the temperature necessary for dishwashers, so most dishwashers come equipped with booster heaters which heat the water an additional time. There are a few common types of commercial dishwashing machines including conveyor belt style machines, single-tank, stationary dishwashers, batch-type dump machines, recirculating non-dump machines, and many more. The best dishwasher for each food service operation will vary depending on their individual needs.

Three-basin sinks should also be of suitable size and should be well-maintained. Sinks should be large enough to clean large pots and pans, dishes, and other equipment and should be kept clean and sanitized regularly. Check for and repair any water leaks immediately.

Service sinks should be used for janitorial tasks and purposes. They are usually located in janitorial areas and used for utility purposes.

Handwashing stations should be readily accessible, including stations in and/or directly adjacent to restrooms. Stations should also be available in food prep areas, food service areas, and near dishwashing areas. All handwashing stations should be clean and well-stocked. This means always having potable hot and cold water, soap, and either paper towels, a continuous towel system, or an air dryer for drying hands. Handwashing stations should not be used for any other purpose, such as storing dirty dishes or other equipment. These stations should not be blocked by equipment or other items.

While it's preferred to have separate restrooms for employees and customers, it isn't required. However, it is required that restrooms must be accessible to customers without requiring them to walk through any food preparation areas. Restrooms should be clean, stocked, and have self-closing doors. Garbage cans must be provided if paper towels are provided, and covered garbage cans must be provided in the women's restrooms.

Building Systems and Utilities

Many systems are necessary to keep facilities operating efficiently and smoothly. Water and plumbing, lighting, ventilation, and garbage collection and disposal systems should all be well-maintained and regularly inspected. The systems must also be sufficient to meet the usage requirements for the building, including food preparation and storage as well as any customer services provided, such as dining rooms and restroom facilities.

Water and Plumbing

All water used in food preparation and in cleaning and sanitizing food-prep surfaces and equipment must be clean and potable, meaning that the water is suitable for drinking. City and local ordinances determine approved water sources, which usually include city-provided public water; private sources such as wells, as long as they are properly maintained and regularly inspected; enclosed, portable water containers; and water trucks. Plumbing should be installed in a way that avoids cross contamination between potable and non-potable water sources. Plumbing should be inspected regularly, and a licensed plumber should be called in immediately upon discovery of any leaks, cracks, or other issues.

One of the biggest causes of water contamination comes from cross connections. A cross connection is when there is a physical link, such as a hose, between clean and dirty water, such as that found in drains or sewers. A hose that is attached to the faucet on a sink with the other ending resting on a drain, for example, is a contamination risk because germs and bacteria can enter the hose from the drain and work its way up through the hose to the clean faucet.

Care should also be taken to avoid cross contamination caused by backflow. Backflow occurs when there is a connection between a clean water source, such as a sink faucet, and a dirty water source, such as the water in a mop bucket. Water carrying contaminants and bacteria from the dirty water source can travel back up into the clean water source when the clean water is turned off, such as when water is siphoned from one source to another, carrying contaminants with it and thus polluting the clean water source. This is known as **backsiphonage**.

Backflow and cross contamination can be prevented by avoiding cross-connections. Hoses should not be attached to faucets unless the other end of the hose is held so that it does not come into contact with any unclean surface. In addition, the use of backflow prevention devices, such as vacuum breakers and reduced pressure zone backflow preventers, can help alleviate the problem. A vacuum breaker is a mechanism that closes a valve, shutting the water supply line when the water is turned off, thus preventing the dirty water from moving back up the hose. Reduced pressure zone backflow preventers have multiple check valves that seal off the water supply. Regardless of type, all backflow prevention devices must be inspected periodically by a trained technician to ensure that they are clean and in good condition. This work should be documented and kept on file by the manager.

The best way to prevent backflow is to allow for an air gap between the clean water and the dirty water. An air gap is a space between the clean water source and the potential contaminants. A properly installed sink should have an air gap between the faucet and the rim of the sink and between the drainpipe and the floor drain. This gap prevents contaminants and bacteria from being able to reach the clean water source at the faucet.

Lighting

Lighting needs vary by area, but lighting should be bright enough that workers can easily see the condition of the foods and the food prep surfaces. Good lighting is also necessary to maintain cleanliness and identify areas that need maintenance. The brightness, or intensity, of lighting is measured in lux, or foot-candles. Requirements, particularly in food prep areas, are dictated by local ordinances. The general minimum lighting requirements are 540 lux for prep areas; 215 lux for handwashing areas, restrooms, buffets, displays, and wait stations; and 108 lux for dining rooms, dry-storage rooms, and inside walk-in freezers and coolers.

Lighting should be inspected regularly. Light sources should use shatter-resistant bulbs of the correct size and luminosity and should be adequately covered to prevent broken glass contamination in the event that a bulb does break. Burnt-out bulbs should be changed immediately as needed.

Ventilation

Ventilation refers to the airflow of the facility. Adequate airflow is necessary to prevent the buildup of heat, smoke, steam, and condensation and to provide employees and customers with fresh air.

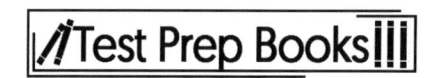

Ventilation also helps to remove any odors or fumes from cooking, cleaning, and/or sanitization. Ventilation systems can collect a buildup from grease and other contaminants, so they should be cleaned regularly. It is also important to have ventilation systems inspected regularly by a licensed technician to ensure that there are no blockages and that the system is in optimal condition.

Garbage

Garbage is a prime breeding ground for bacteria, and it is attractive to bugs, rodents, and other pests. Indoor garbage cans and containers should be waterproof, leakproof, and pest proof and should remain covered when not in use. Garbage should be removed from food prep areas as quickly as possible, and care should be taken to avoid contaminating the food prep areas in the process. Garbage cans should be cleaned regularly to avoid the buildup of contaminants that could cause illness, including wiping down the outside of the cans. Garbage cans should always be removed from the food prep areas for cleaning.

Garbage storage areas should also be well maintained. Indoors, garbage and recyclables should be kept in a designated storage area away from food storage and food prep areas. Outdoor areas should be kept clean, and storage containers, such as dumpsters, should be enclosed to prevent pest infestations. Outdoor garbage areas should also allow for suitable draining in the event of spills and rain.

Maintenance

The key to a clean, well-functioning facility is good maintenance practices. The entire facility should be cleaned on a regular basis. All building systems should be inspected regularly and in accordance with a regular maintenance schedule. The building should be inspected for cracks, holes, and other damaged areas that could allow for leaks and pest infestations. Pest control measures should be implemented and completed regularly. Finally, outdoor areas, including patios, parking lots, and walkways, should be kept clean, well-lit, and free of hazards.

Practice Test #1

1. Which of the following situations risks cross-contamination of food?
    a. Using a knife to cut poultry, and then slice pork
    b. Rotating packaged meat with bare hands
    c. Thawing raw beef on the bottom shelf of a reach-in fridge
    d. Mixing lettuce, carrots, and celery for salad with gloved hands

2. Which of the following is an alternative to elevating floor- or table-mounted equipment?
    a. Setting directly on the floor without mounting so that it can be moved easily
    b. Putting the equipment on rubber feet to prevent sliding or movement
    c. Affixing and sealing the equipment to the surface
    d. None; the equipment must be elevated in all cases

3. What is the minimum internal temperature that ground beef must be cooked to in order to be safe for consumption?
    a. 165°F for 17 seconds
    b. 145°F for 17 seconds
    c. 160°F for 17 seconds
    d. 155°F for 17 seconds

4. What temperature should the water be when mixing a chemical sanitizer that contains iodine?
    a. 50°F
    b. 100°F
    c. 75°F
    d. 68°F

5. What temperature range listed is the *temperature danger zone*?
    a. 41°F – 135°F
    b. 41°F – 138°F
    c. 40°F – 140°F
    d. 45°F – 140°F

6. Which of the following is NOT a suitable hand-drying option in handwashing stations?
    a. Clean hand towel
    b. Paper towels
    c. Continuous towel system
    d. Air dryer

7. What is the accrediting agency for NSF?
    a. FDA
    b. ANSI
    c. USDA
    d. CDC

8. How should egg dishes be prepared to be safe for consumption if they require the egg to be not fully cooked or raw?
   a. Egg dishes that require the egg to be raw or not fully cooked should be cooked to at least 145°F.
   b. Dishes that require eggs to be not fully cooked or raw should be prepared with eggs treated for pasteurization.
   c. Egg dishes with raw egg should be cooked until the yolks are firm, though the whites can be runny.
   d. There are no special considerations that need to be made for egg dishes with raw egg since eggs are not a risk for illness.

9. What are the proper steps for cleansing and sanitizing a surface that will be used for food?
   a. Remove excess food, clean the surface with cleaning solution, rinse, apply sanitizing solution with a reusable cloth, and air dry for at least one minute.
   b. Remove excess food, clean the surface with cleaning solution, rinse, apply sanitizing solution with a disposable wipe, and air dry for at least one minute.
   c. Clean the surface with a cleaning solution that removes excess food, rinse, apply sanitizing solution with a disposable wipe, and air dry for at least one minute.
   d. Remove excess food, clean the surface with cleaning solution, rinse, apply sanitizing solution with a disposable wipe, and wipe dry.

10. Which of the following is NOT a requirement for the construction material for food service equipment?
    a. Corrosion resistant
    b. Nonabsorbent
    c. Stainless steel
    d. Smooth

11. Which of the following groups is NOT considered a highly susceptible population (HSP)?
    a. Children under five
    b. People experiencing poverty
    c. People with kidney disease
    d. Older adults

12. Which of the following is the safest way to thaw frozen meat?
    a. Setting the frozen meat down on a counter to allow it to thaw to room temperature
    b. Placing the frozen meat in a sink full of hot water
    c. Thawing frozen meat in the microwave and then placing it in the refrigerator for later use
    d. Moving the frozen meat into the refrigerator

13. Meredith needs to sanitize the restaurant's utensils after the lunch rush. What is the best method to use?
    a. Clean them in soapy water and rinse well.
    b. Soak them in very hot water.
    c. Spray them with a quaternary ammonium compound.
    d. Wipe them down with a clean towel.

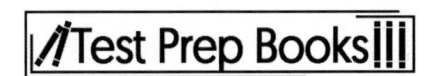

14. Which statement is most accurate regarding food contamination?
    a. Food that is properly wrapped or stored in a container is unlikely to become contaminated.
    b. Once food is placed in a buffet serving area, food employees have no control over the risk of contamination.
    c. Food employees are unable to control risks from contamination that occurs before food arrives at the facility.
    d. The risk of contamination can be decreased by sending sick employees home.

15. The Sunday Brunch Spot is having their kitchen remodeled. The construction contractor has recommended using a textured spray on the walls to help hide any defects in the new drywall before painting, but per the owner's request, he has made several options available. Which option would be the best choice?
    a. Go with the textured spray; the contractor is the expert here.
    b. Choose a smooth, durable surface, such as wall tiles.
    c. Do the textured spray, but make sure the paint is washable.
    d. Request a smooth drywall finish with durable, washable paint.

16. Aflatoxin is commonly found in which of these foods?
    a. Fish
    b. Tomatoes
    c. Nuts
    d. Eggs

17. How often should machines that hold and/or dispense TCS foods be cleaned?
    a. Twice daily
    b. Daily
    c. Twice weekly
    d. Weekly

18. If a can of green beans is at the room temperature of 67°F before and after opening, what is the maximum amount of time the green beans can remain at this room temperature before being consumed or discarded?
    a. 2 hours
    b. 4 hours
    c. 6 hours
    d. They cannot be held at this temperature.

19. What is the level of acidity or alkalinity in the water called?
    a. pH
    b. Hardness
    c. Buffering capacity
    d. Salinity

20. Which of the following is NOT a time when food contact surfaces should be cleaned and sanitized?
    a. After use
    b. When changing to prepare a new kind of food
    c. Whenever a task is interrupted and contamination may have occurred
    d. Every six hours during continuous use

Practice Test #1

21. Which term refers to the removal of dirt and debris, such as food particles, from a surface?
    a. Sanitizing
    b. Disinfecting
    c. Cleaning
    d. Polishing

22. Which method would be most effective for preventing Scromboid poisoning?
    a. Ensure fish are kept at or below 41°F beginning as soon as possible after being caught or harvested.
    b. Ensure fish are not caught from bodies of water with high histamine concentrations.
    c. Cook fish to an internal temperature of at least 145°F for at least 15 seconds.
    d. Avoid fish with an unusual appearance, smell, or taste.

23. A kitchen is cooking ground beef in preparation for batch cooking a casserole on tomorrow's menu. The employee cooks the beef in a microwave oven to an internal temperature of 150°F, then begins to let it cool in shallow pans. The ground beef finished cooking at 4:00 PM. The ground beef must be cooled to which temperature or below at 6:00 PM?
    a. 100°F
    b. 80°F
    c. 41°F
    d. 70°F

24. What is the general prescribed time range to allow a chemical sanitizer that contains chlorine to remain on a surface?
    a. At least seven seconds
    b. At least sixty seconds
    c. At least thirty seconds
    d. At least ten seconds

25. How should the strength of chemical sanitizer be checked?
    a. Measuring the temperature
    b. Reading the label on the package
    c. Using a test kit
    d. Mixing one part chemical to one part water

26. What causes glassware to experience thermal shock?
    a. Glass moving quickly from very cold to very hot
    b. Glasses being stacked on top of one another
    c. Glasses hitting each other while soaking in a sink
    d. A glass reaching room temperature before being run in the dishwasher

27. Under what conditions would it be acceptable to use a cutting board to cut fish after using it to cut chicken?
    a. When the cutting board is turned over after cutting the chicken so that the clean side is used for the fish
    b. When a separate set of single-use gloves is used with each item
    c. When the chicken and fish will be combined for cooking as part of the same dish
    d. When there is no risk that the customer ordering the fish is allergic to chicken

28. What is the minimum temperature that a brined ham must be cooked to in order to be safe for consumption?
   a. 140°F
   b. 145°F
   c. 155°F
   d. 160°F

29. Jane is a manager at the Silver House restaurant. Lately, the restaurant's produce supplier has not been meeting its needs, and Jane would like to change suppliers. One of the waitstaff has a cousin who is a farmer and farm-to-table supplier who might be able to take over. Who can Jane contact to see if this new supplier has undergone inspections?
   a. FDA or HACCP
   b. FDA or USDA
   c. USDA or HACCP
   d. HACCP or GMP

30. What information should be included in a supplier's safety inspection report?
   a. Receiving, storage, and processing practices
   b. Staff training and diversity
   c. Types of food products provided by the supplier
   d. Geographical range of the supplier's delivery route

31. When washing dishes in a three-compartment sink, what is the minimum temperature that the water should be in the first (washing) sink?
   a. 100°F
   b. 110°F
   c. 105°F
   d. 115°F

32. What is the minimum temperature at which hot food may be stored ready-to-serve?
   a. 135°F
   b. 140°F
   c. 165°F
   d. 90°F

33. At 4:30 PM, the manager checks how long ingredients on their restaurant's salad tray have been sitting out, and their current temperature.

Ingredient	Time Set Out	Current Temperature
Lettuce	12:30 PM	75°F
Cucumber	10:00 AM	60°F
Carrots	2:00 PM	45°F

They should remove and discard which ingredients?
a. Lettuce
b. Lettuce, cucumber
c. Cucumber, carrots
d. Lettuce, carrots, cucumber

34. What is the term used for disease-causing microorganisms such as bacteria, viruses, parasites, and fungi?
a. Pathogens
b. Infections
c. Contaminants
d. Diagnoses

35. What is the industry-standard measurement used to determine the appropriate strength of sanitizer to water?
a. mg/L
b. ppth
c. % m/m
d. ppm

36. Cold-held food should be held at or below what temperature?
a. 38°F
b. 41°F
c. 42°F
d. 44°F

37. Alice is cleaning the food prep countertops in her restaurant. What does she need to do this job?
a. Soap and water
b. Sanitizing chemicals
c. A wet rag
d. A mop bucket

38. When receiving a delivery, a visual inspection should include which of the following?
a. The purchase order
b. The delivery driver
c. The outside of the delivery truck
d. The inside of the delivery truck

39. Which of the following is the proper way to store food?
    a. Foods can be held within the temperature danger zone if they are not held there for too long.
    b. Old foods should be stored behind newer foods.
    c. So long as foods are kept out of the temperature danger zone, there is no specific way they need to be stored.
    d. Raw foods should be stored away from and below ready-to-eat foods to prevent cross-contamination.

40. What is the proper way to hold food that is being marinated?
    a. When food is being marinated, it should be held in the fridge.
    b. Food being marinated can be held at or below room temperature.
    c. Food should only be marinated in the freezer.
    d. If a marinade has been boiled, the temperature it is held at does not matter.

41. Which of the following is one of the CDC's five foodborne illness risk factors?
    a. Inadequate employee training
    b. Improper hand washing
    c. Improper holding times and temperatures
    d. Fecal contamination

42. How should fresh meat and poultry be temperature checked?
    a. By inserting a thermometer directly into the thickest part of the meat
    b. By inserting a thermometer directly into the thinnest part of the meat
    c. By placing the thermometer between two pieces of meat
    d. By folding the meat around the thermometer

43. What are the key aspects of using chemical sanitizers?
    a. Concentration, temperature, pH, water hardness, and time
    b. Chemical, temperature, pH, water amount, and time
    c. Concentration, temperature, pH, water amount, and total application
    d. Concentration, time, preparation, water hardness, and total application

44. Tags for shellfish must be held on file for how many days after the last shellfish is used?
    a. 70 days
    b. 45 days
    c. 90 days
    d. 30 days

45. At what temperature should cold foods, such as meats, be delivered?
    a. 35°F or cooler
    b. 41°F or cooler
    c. 45°F or cooler
    d. 47°F or cooler

46. Documents for any fish that will be used for raw dishes should include information on which of the following?
    a. What the fish was fed
    b. How the fish was stored
    c. The color of the fish
    d. The length of the fish

47. Which list includes each of the major food allergens?
    a. Meat, poultry, eggs, raw milk products, seafood, and raw vegetables, grains, and fruits
    b. Peanuts, tree nuts, sesame, eggs, crustacean shellfish, wheat, milk, soybeans, and fish
    c. Gluten, latex, leafy greens, sprouts, raw flour, dairy, nuts, seeds, and refined sugar
    d. Tuna fish, tofu, cashew nuts, deviled eggs, lobster, couscous, soft cheeses, and hummus

48. What are the steps to properly cleaning a surface?
    a. Wipe surface with a towel, wash with soapy water, rinse, dry
    b. Remove debris and stuck-on materials, scrub with soapy water, rinse, dry
    c. Remove debris and food particles, scrub with soapy water, rinse
    d. Remove debris and stuck-on materials, scrub with soapy water, dry

49. Which of the following is NOT an example of a TCS food?
    a. Milk
    b. Cut melon
    c. Eggs
    d. Uncooked beans

50. Which of these is an example of physical contamination?
    a. Particles of insulation that fell into an open bag of oats from a ceiling repair
    b. *Shigella* bacteria from a food worker who did not practice proper glove use contaminating pasta salad
    c. High levels of mercury in rice that was treated with a mercury-based fungicide
    d. Ciguatoxin present in snapper caught in Florida

51. Toxoplasmosis would best be classified as which of the following?
    a. A foodborne infection
    b. A physical hazard
    c. A rare condition
    d. A developing-world illness

52. Which of the following is an example of a sanitary personal hygiene practice?
    a. Washing hands in warm soapy water for at least twenty seconds
    b. An employee handling food wearing a few rings on their hands
    c. Hair not being contained by a hair restraint while preparing food
    d. An employee keeping their food near where food is being prepared for customers

53. If a customer has food allergies and has requested that their meal be free of the nine major food allergens, which action would be recommended for protecting the customer?
    a. Rinse off uncooked chicken that was stored with shrimp and fish before cooking it for the customer.
    b. Prepare the meal in a separate area with freshly cleaned and sanitized equipment.
    c. Ensure that eggs and milk products used as ingredients are pasteurized and thoroughly cooked.
    d. Pick out all nuts on a premade salad before serving it to the customer.

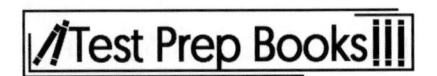

54. Which of these symptoms would most likely indicate that a customer is having a food allergy reaction?
    a. Jaundice
    b. Tingling around the mouth
    c. Difficulty breathing
    d. Weight loss

55. What creates a cross connection with regards to water and plumbing?
    a. The space between the faucet and the edge of the sink
    b. A hose attached to the sink with the other end in a mop bucket
    c. The space between the bottom drain of the sink and the floor
    d. A sink where the faucet is also a handheld sprayer

56. What is backflow?
    a. When dirty water travels back up a hose to a clean water source
    b. When air and odors are drawn up from the food prep areas through the ventilation system
    c. When swinging kitchen doors allow odors to escape into the dining areas and other parts of the building
    d. When bugs and pests enter a building through the fly fans

57. Which building system prevents the build-up of heat, smoke, steam, and condensation?
    a. HVAC
    b. Ventilation
    c. Plumbing
    d. Electrical

58. What minimum internal temperature must eggs that will be hot-held for service be cooked to?
    a. 145°F
    b. 150°F
    c. 155°F
    d. 160°F

59. Which of the following foods can safely be re-served to a customer?
    a. There are no safe foods for re-service
    b. A dish of lemons that were not used for drinks
    c. An unopened, single-serve ketchup packet
    d. An untouched side salad

60. Michelle accepted an order of shucked shellfish that was at a temperature of 44°F. The driver told her he had been on the road for three hours already that morning. Michelle put the shellfish in the refrigerator within an hour. Five hours later, she checked the shellfish and found the temperature to be 42°F. Can the chefs begin using this shellfish for the evening menu?
    a. Yes, the shellfish was delivered at an appropriate temperature.
    b. Yes, the shellfish was put into the refrigerator within an hour of delivery.
    c. No, the shellfish did not cool to a sufficient temperature within the prescribed time frame.
    d. No, the shellfish should not have been on the truck for so long.

Practice Test #1

61. Which of the following is a sign that a frozen food is potentially unsafe to keep for consumption due to thawing and refreezing upon arriving in a shipment?
    a. The packaging is broken open.
    b. There are ice crystals on the packaging.
    c. The packaging is bloated.
    d. The frozen food is discolored.

62. Concerning bacterial contamination, which of these is a true statement?
    a. Bacteria grow only in the danger zone.
    b. Certain bacteria produce toxins.
    c. Proper cooking will kill all bacteria.
    d. Bacterial infections result in immediate symptoms.

63. Foods should be cooled to 70°F within two hours and 41°F within four hours. Methods to speed the cooling of larger batches of food do NOT include:
    a. Stirring the food in a container placed in an ice bath
    b. Using rapid cooling equipment
    c. Using a proofing cabinet
    d. Maximizing the exposed surface area of the food

64. If a pan of lasagna is removed from cooking at 375°F and must be cooled and refrigerated, what is the maximum amount of time the lasagna itself, as measured by a food thermometer, can safely remain in the danger zone temperature range?
    a. 2 hours
    b. 4 hours
    c. 6 hours
    d. 8 hours

65. Which of the following is the proper way to prepare produce for consumption?
    a. Inspect, throw out rotten produce and remove damaged parts, rinse thoroughly (even if the skin or rind is not eaten), cut
    b. Inspect, throw out rotten produce and remove damaged parts, rinse thoroughly (unless the skin or rind is not going to be eaten), cut
    c. Rinse thoroughly (even if skin or rind is not eaten), cut, throw out rotten produce, remove damaged parts
    d. Inspect, throw out rotten produce and remove damaged parts, rinse thoroughly with soap (even if the skin or rind is not eaten), cut

66. Where should handwashing stations be located?
    a. In dining rooms, restrooms, food prep areas, and food services areas, dishwashing areas
    b. In or near restrooms, food prep areas, and food service areas, dishwashing areas
    c. In dining rooms, behind the bar area, and in food prep and service areas, dishwashing areas
    d. In restrooms, bar areas, food prep areas, and service areas

67. Which of these is the leading cause of foodborne illness in the United States?
    a. *E. coli*
    b. *Salmonella*
    c. Mold
    d. Norovirus

68. Hepatitis A is an example of what type of pathogen?
    a. Bacterium
    b. Parasite
    c. Virus
    d. Fungus

69. What is the minimum internal temperature that raw poultry must be cooked to?
    a. 145°F
    b. 150°F
    c. 160°F
    d. 165°F

70. Ready-to-eat TCS foods that will not be sold or consumed within twenty-four hours must be clearly marked with the date by which they should be consumed or sold. How many days after preparation is the expiration date for refrigerated ready-to-eat TCS foods?
    a. Three days
    b. Five days
    c. Seven days
    d. Nine days

71. Which term best describes an illness caused by the presence of too much copper in spaghetti sauce that was stored in a copper pan?
    a. Foodborne infection
    b. Foodborne intoxication
    c. Foodborne illness due to physical contamination
    d. Scombroid poisoning

72. A pork roast must be cooked to what minimum internal temperature?
    a. 140°F
    b. 145°F
    c. 150°F
    d. 155°F

73. Belinda's manager has asked her to receive the order of meat from the supplier. What equipment and information does Belinda need to have to properly receive and inspect the delivery?
    a. Purchase order and thermometer
    b. Thermometer and scale
    c. Purchase order and tape measure
    d. Tape measure and scale

74. Which of the following is true concerning proper food storage methods used to prevent cross-contamination?
    a. Ready-to-eat food should be stored above raw food.
    b. Foods should be held at temperatures between 41°F and 135°F.
    c. Foods should be used last in, first out.
    d. Uncooked meat should be kept in a separate cooler from produce.

75. Manuel has a new dry goods supplier who would like to schedule a delivery for before the restaurant opens in the morning, when no staff or management would be available to receive the delivery. Should Manuel permit this?
    a. No, management or staff should always be available to receive the delivery.
    b. Yes, this is called a key drop delivery and is fine.
    c. No, a key drop delivery should only be permitted by well-known, trusted suppliers.
    d. Yes, the delivery driver can leave the goods just outside the back door of the restaurant.

76. How is the intensity of lighting measured?
    a. Lumens
    b. Luminous flux
    c. Ohms
    d. Lux

77. Which of the following is true of the use of a dishwasher in cleaning and sanitizing dishes?
    a. All dishes that are used in preparing food can go in the dishwashers used in professional establishments.
    b. High-temperature dishwashers are unable to be used for glassware.
    c. Both chemical and high-temperature dishwashers can clean and sanitize dishes.
    d. The dishwasher should be used for sanitizing dishes, while the three-compartment sink should be used for cleaning dishes.

78. What is the proper temperature for hot-held food to be kept at?
    a. 140°F and above
    b. 135°F and above
    c. 138°F and above
    d. 145°F and above

79. Which federal regulatory agency sets forth the cGMPs?
    a. USDA
    b. CDFA
    c. NIH
    d. FDA

80. A shipment of food is received. The temperature of the cheese is 43°F, the temperature of the uncooked rice is also 43°F, and the raw frozen chicken is received with ice crystals. Which of these foods is safe to keep for consumption?
    a. The cheese
    b. The chicken
    c. The rice
    d. None of the food is safe to be kept.

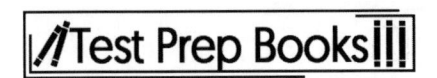

81. What is one step that can be taken to address one of the five most common foodborne illness risk factors?
    a. Conduct periodic visual inspections of the interior of randomly selected raw animal products to ensure they are not contaminated with toxins or pathogens.
    b. Post signage in dining areas warning customers of the dangers of sharing utensils, cups, and straws.
    c. Train employees on proper hand washing and its importance for preventing foodborne illnesses.
    d. Avoid using equipment that contains glass that could break, such as glass thermometers that are not covered with a shatterproof coating.

82. Other than temperature, what is one way to tell that eggs have been fully cooked?
    a. The whites and yolks are both firm.
    b. The whites are firm.
    c. The yolks are firm.
    d. The whites and yolks are partially firm.

83. Which of the following is the proper practice to replace old food in a self-service bar?
    a. The container of old food should be replaced with an entirely new container of food, which should be covered as it is brought from the kitchen to the self-serve area.
    b. So long as the temperature is held above 135°F, new food can be added to old food after being brought out from the kitchen to the self-serve area.
    c. The container of old food should be removed and replaced with a container of new food and does not need to be covered on its way to the self-serve area.
    d. Old food should be removed from the container, after which new food can be placed into the container.

84. Which type of shelving is best for coolers and freezers?
    a. Wire
    b. Glass
    c. Plastic
    d. Metal

85. Anthony has noticed that racoons seem to be getting into the outside garbage containers at his facility. What can he do to prevent this?
    a. Improve the lighting in that area.
    b. Take garbage outside less often.
    c. Improve the drainage to prevent liquids from pooling.
    d. Build an enclosure, such as a fence or wall, around the outside garbage areas.

86. At what temperature should hot foods be delivered?
    a. 130°F
    b. 133°F
    c. 135°F
    d. 137°F

87. Marcus has recently taken over as the food safety manager at a restaurant. He has discovered that the manufacturing guidelines for one type of product have not been reviewed or updated in quite some time. This could be a violation of which of the GMPs?
    a. Products
    b. Processes
    c. Procedures
    d. People

88. Which of the following is NOT true when it comes to handling glassware?
    a. Glassware should be allowed time to warm after ice is removed and before being washed.
    b. Glassware should be allowed time to cool after being removed from the dishwasher before adding ice or cold beverages.
    c. Glassware should not be allowed to rest in the dishwater while being washed.
    d. Glassware should be stacked to enable an employee to carry multiples safely.

89. Which of the following are part of a building's systems and utilities?
    a. Plumbing, lighting, ventilation, and waste disposal
    b. Floors, walls, ceilings, windows, and doors
    c. Coolers and freezers, counters, and storage shelving
    d. Ventilation, floors and walls, coolers and freezers, and waste disposal

90. Which of the following is a TCS food?
    a. Uncooked pasta
    b. Whole, raw bell pepper
    c. Whole, raw melon
    d. Cooked onions

Answer Explanations #1

1. A: Choice A is correct because utensils used to handle raw meat should be washed before handling another type of raw meat. This prevents the spread of bacteria from one type of meat to the other. Choice B is incorrect because sealed meat packaging is not considered a contaminant; only opened packaging. Choice C is incorrect because placing meat on the bottom shelf appropriately avoids dripping onto other foods. Choice D is incorrect because the employee is wearing gloves.

2. C: Floor- and table-mounted equipment can be affixed directly to the surface and sealed instead of elevating it. Putting the equipment directly on the floor, Choice A, or on rubber feet, Choice B, is not sufficient. While elevating the equipment is common, it is not the only option, as in Choice D.

3. D: Choice D correctly states that the minimum internal temperature that ground beef must be cooked to for safety is 155°F for at least 17 seconds. Choices A, B, and C are incorrect as 165°F, 145°F, and 160°F are not the minimum internal temperature that ground beef must reach.

4. D: When mixing an iodine sanitizer, the optimal water temperature is 68°F. Choices A, B, and C are incorrect.

5. A: Choice A states the correct temperature danger zone: 41°F–135°F. Choices B, C, and D are incorrect because they do not list the correct temperature ranges of the temperature danger zone.

6. A: Reusable towels, such as traditional hand towels, can hold bacteria and are not suitable for commercial handwashing stations. Choices B, C, and D are better options.

7. B: NSF is overseen by the American National Standards Institute (ANSI). The FDA, Choice A, and the USDA, Choice C, are both regulatory agencies for food safety. The CDC, Choice D, is the Centers for Disease Control.

8. B: Dishes that require raw or not fully cooked eggs should be prepared with eggs that have been pasteurized. Choice A is incorrect as there is not a temperature that these dishes must be cooked to; rather, they should be made with treated eggs. Choice C is incorrect as in egg dishes with raw egg, neither part of the egg will be firm. Also, pasteurized eggs should be used in this scenario. Choice D is incorrect as eggs are a risk for illness, and there are considerations to make for dishes that use raw or not fully cooked eggs.

9. B: Choice B states the correct steps. First, the surface is cleared of excess food before a cleaning solution is used to clean the surface. The cleaning solution is then rinsed, and the sanitizing solution is applied, which then air dries for at least one minute. Choice A is incorrect because it states that the sanitizing solution should be applied with a reusable cloth rather than a disposable one. Choice C is incorrect because it states that the cleaning solution removes the excess food. However, excess food should be removed before the cleaning solution is used. Choice D is incorrect; it states that the surface should be wiped dry after using the sanitizing solution, not air dried.

10. C: While stainless steel is not a requirement for food service equipment, it is a popular choice because it is corrosion resistant, Choice A, nonabsorbent, Choice B, and smooth, Choice D.

11. B: Apart from other risk factors that would classify them as part of a highly susceptible population (HSP), people experiencing poverty are not considered an HSP. Choice A is incorrect because children

Answer Explanations #1

under five are considered preschool-aged children and are an HSP because their immune systems have not yet fully developed. Choice C is incorrect because people who have kidney disease or other diseases that weaken the immune system are considered immunocompromised people, making them part of that HSP. Choice D is incorrect because older adults are considered an HSP.

12. D: Choice D states one of the safe ways to thaw food of the three provided in the text. Choice A is incorrect as food should not be left to thaw at room temperature. Choice B is incorrect as food should be thawed in cold water, not hot water. Choice C is incorrect; while food can be thawed in the microwave, it is safest if it is cooked immediately following thawing.

13. B: Heat sanitization, submerging the utensils in water that is at least 171°F and soaking them for at least thirty seconds, is the best choice here. Cleaning the utensils with soapy water, Choice A, is good to remove any food debris but will not sanitize them. Using a chemical sanitizer, such as a quaternary ammonium compound in Choice C, is good for food prep surfaces, but it is not an ideal choice for small items like utensils. Simply wiping with a towel, Choice D, will not sufficiently clean nor sanitize the utensils.

14. D: Sick employees are more likely to be contagious and contaminate food with pathogens, so excluding them from work will help reduce the risk of biological contamination of food. Choice A is incorrect because wrapping or enclosing time/temperature control for safety (TCS) foods will not protect them from becoming contaminated by overgrowth of pathogens through time and temperature abuse. Choice B is incorrect because there are still important steps that food employees can and should take to control the risk of contamination, such as ensuring food is shielded from overhead and maintained at a proper temperature and providing appropriate utensils to serve food and clean plates for customers making return trips to the buffet. Choice C is incorrect because there are important steps that food employees can and should take to control risks associated with contamination that occurs before food arrives at a facility, including ensuring that food is only received from approved facilities and that it is received at a proper temperature.

15. B: Walls in food service facilities should be smooth, durable, and easy to clean. In the case of a kitchen area, wall tiles would be the best choice. Choices A, C, and D all allow for a more porous surface, which can harbor bacteria.

16. C: Aflatoxin is a mycotoxin produced by some types of *Aspergillus* molds. It grows naturally on some crops, especially grains and nuts, and can be found in the milk of cows that eat contaminated grains. Choices A, B, and D are incorrect because fish, tomatoes, and eggs are not as likely to be contaminated with aflatoxin.

17. B: Machines that hold and/or dispense TCS foods should be cleaned daily according to the manufacturer's directions and local regulations. Choices A, C, and D are incorrect.

18. B: Because the temperature of the green beans is at or below 70°F when they become a TCS food (when the can is opened) and during the holding time, the maximum amount of time they can be held without temperature control is 4 hours. Choice A is incorrect because, as long as their temperature remains at or below 70°F, the green beans do not need to be used or discarded this quickly. Choice C is incorrect because 6 hours is too long to hold a TCS food without temperature control unless the food starts out at or below 41°F and remains at or below 70°F for the duration of the time. Choice D is incorrect because canned green beans can be held at 67°F as long as they are properly labeled and consumed or discarded within 4 hours.

19. A: The level of acidity in water is its pH level. Water hardness, Choice *B*, refers to the level of minerals in the water. Buffering capacity, Choice *C*, is the water's ability to maintain a stable pH. Choice *D* refers to the number of dissolved substances in the water.

20. D: Food contact surfaces should be cleaned and sanitized every four hours during continuous use, as well as at all the times listed in Choices *A*, *B*, and *C*.

21. C: *Cleaning* is the removal of dirt and debris from surfaces. *Sanitizing*, Choice *A*, is the removal of bacteria, and *disinfecting*, Choice *B*, removes bacteria and viruses. *Polishing*, Choice *D*, generally refers to the appearance of a surface rather than its cleanliness.

22. A: Scromboid poisoning is an overdevelopment of histamine that occurs after fish are harvested and can be prevented by holding fish under refrigeration temperatures. Choice *B* is incorrect because, unlike other fish toxins, the histamine toxin does not accumulate from the fish's environment and food sources. Choice *C* is incorrect because, although it is important to cook fish to an internal temperature of 145°F for at least 15 seconds, freezing and cooking do not inactivate fish toxins. Choice *D* is incorrect because, although histamine toxins can sometimes cause the fish to have an unusual appearance, smell, or taste, this is not a reliable indicator of whether the toxins are present.

23. D: Choice *D* is correct because hot foods which are being cooled must be reduced to 70°F or less within two hours of being cooked. Thus, Choices *A* and *B* are incorrect. Choice *C* is incorrect because the ground beef must be cooled to 41°F after an *additional* four hours—not within four hours of the end of cooking.

24. C: The manufacturer's instructions will give specific information, but generally, sanitizer that contains chlorine should remain on the surface for at least thirty seconds. Choices *A*, *B*, and *D* are incorrect.

25. C: The concentration of chemical sanitizer should be checked using a test kit, which is often included with the product. The test kit helps ensure that the mixture contains the right amount of chemical mixed with the right amount of water. While the temperature of the water, Choice *A*, is important, that does not help to measure concentration. Similarly, the label on the package, Choice *B*, will instruct as to the appropriate concentration to use, but reading the label will not measure the mixture itself. The appropriate mixture should be prepared according to the label instructions, not simply by using equal measures as in Choice *D*.

26. A: Choice *A* is correct as thermal shock is caused when a glass moves too quickly from one extreme temperature to another. Choice *B* is incorrect; while glasses should not be stacked on top of one another, this is not a cause of thermal shock. Choice *C* is incorrect; glasses should not be allowed to make contact in this way, but glasses hitting one another while soaking in a sink is not a cause of thermal shock. Choice *D* is incorrect because glass should be allowed to reach room temperature before being run in the dishwasher; this prevents thermal shock from occurring.

27. C: Because the cooking temperature for fish is lower than the cooking temperature for chicken, no equipment or utensils should be used with fish after being used with raw chicken without first being washed and sanitized unless the fish and chicken will be cooked together as part of the same dish. Choice *A* is incorrect because turning over a cutting board does not prevent cross-contamination. Choice *B* is incorrect because cross-contamination would occur from the cutting board regardless of whether clean gloves are used with the fish. Choice *D* is incorrect because the risk of cross-contamination from

Answer Explanations #1

chicken to fish involves pathogens that can affect anyone, regardless of whether they have food allergies.

28. C: Choice *C* is correct, as mechanically tenderized meat like a brined ham must be cooked to a minimum temperature of 155°F for at least 17 seconds. Choices *A, B, and D* list incorrect cooking temperatures for a brined ham.

29. B: The inspection reports for suppliers can be found by contacting either the US Food and Drug Administration (FDA) or the US Department of Agriculture (USDA). The HACCP, the Hazard Analysis Critical Control Points, Choices *A* and *C*, is a food safety management system, and GMPs, Choice *D*, are Good Manufacturing Practices.

30. A: A safety inspection report should include the supplier's receiving, storage, and processing practices. While staff training should be included, Choice *B*, the diversity of the staff is irrelevant to a safety inspection report. The report does not typically include information about specific products, Choice *C*, or delivery routes, Choice *D*.

31. B: Choice *B* is correct as the minimum temperature that the water of the first, or washing, compartment of the three-compartment sink should be is 110°F. The other choices are incorrect as they are not the minimum temperature that the water of a three-compartment sink must be.

32. A: Ready-to-serve hot food may be stored at a consistent internal temperature of 135°F for up to four hours. Thus, Choice *A* is correct and Choice *D* is incorrect. Choice *B* is incorrect because ready-to-serve hot food may be stored at just below the upper end of the danger zone (140°F). Choice *C* is incorrect because 165°F is the required internal temperature for completing cooking of food.

33. B: Ready-to-eat cold food must be kept below 70°F, so they must remove the lettuce. Such food must also be removed if it has been out for six hours or longer, so the manager must remove the cucumber. The carrots are safe because they have been out for two and a half hours, and their current temperature is below 70°F. Thus Choice *B* is correct, and Choices *A, C,* and *D* are incorrect.

34. A: Pathogens are microorganisms that cause disease. Choice *B* is incorrect because the word *infection* refers to the process in which microorganisms enter the body. It would be correct to say that a pathogen has infected a person or that a person has become infected with a pathogen. Choice *C* is incorrect because *contaminant* is a more general term that can refer both to microorganisms—which fall into the category of biological contaminants—and other types of contaminants, including toxins, chemical contaminants, and physical contaminants. Choice *D* is incorrect because a diagnosis is the identification of a disease. While the name of an individual diagnosis may be the same as, or based on, the name of the pathogen that caused it (for example, a diagnosis of salmonellosis or hepatitis A), the term *diagnoses* does not refer to disease-causing microorganisms in general.

35. D: Concentration is measured in parts per million (ppm), which should be checked using a test kit. Choice *A* is milligrams per liter. Choice *B* is parts per thousand, and Choice *C* is percentage composition by mass. These are all different ways of measuring concentration, but the industry standard is to use ppm.

36. C: Choice *B* states the correct temperature that cold-held food should be held below, 41°F. Choice *A* is incorrect because 38°F is too low. Choices *C* and *D* are incorrect as the temperatures 42°F and 44°F are too high and are in the temperature danger zone.

37. B: Food prep surfaces should be sanitized using specific chemicals designed to kill germs. Cleaning with soap and water, Choice A, is fine for walls and floors. Rags, Choice C, should not be used for cleaning, and a bucket used for mopping floors, Choice D, should never be used to clean food prep areas.

38. D: Receipt of a delivery should include a visual review of the truck to look for any signs of contamination, such as bug infestation or rodent droppings. Examining the outside of the truck, Choice C, is not necessary, nor is a judgement about the visual appearance of the delivery driver, Choice B. The purchase order, Choice A, should have been reviewed upon placement of the order, rather than when the delivery has arrived.

39. D: Choice D states that raw foods should be stored away from and below ready-to-eat foods to prevent cross-contamination. Choice A is incorrect as storing food in the temperature danger zone should generally be avoided as it encourages bacterial growth. Choice B is incorrect as storing older foods behind newer foods can allow older foods to go bad before being used. Choice C is incorrect; while food should be kept out of the temperature danger zone, there are other things to keep in mind when storing food (such as avoiding cross-contamination).

40. A: Choice A is correct; when food is being marinated, it should be refrigerated. Choice B is incorrect because this could lead to food being stored within the temperature danger zone. Choice C is incorrect; while marinated food can be frozen, it is not the only way to hold marinated food. Choice D is incorrect because while boiling a marinade may kill germs within it, it does not make the temperature that it's held at irrelevant.

41. C: The five foodborne illness risk factors are food from unsafe sources, inadequate cooking, improper holding times and temperatures, contaminated equipment, and poor personal hygiene. Choice A is incorrect because, although employee training does impact each of the five foodborne illness risk factors, it is not specifically identified by the CDC as one of those five factors. Choice B is incorrect because, although improper hand washing does contribute significantly to foodborne illnesses, in reference to the five foodborne illness risk factors, hand washing falls into the broader category of poor personal hygiene. Choice D is incorrect because, although fecal contamination does contribute significantly to foodborne illnesses and is strongly related to poor personal hygiene, food from unsafe sources, and contaminated equipment, it is not itself one of the CDC's five foodborne illness risk factors.

42. A: Fresh meat and poultry should be checked by placing a thermometer directly into the thickest part of the meat. The thinnest part of the meat, Choice B, would not give an accurate reading for larger sections or pieces of meat. Choice C, placing a thermometer between two packages of meat, or by folding the package around the thermometer, Choice D, is used for meat that is vacuum-sealed.

43. A: The five key aspects of using chemical sanitizers are concentration, temperature, pH, water hardness, and time. Choices B, C, and D are incorrect.

44. C: Tags for shellfish must be held on file for at least 90 days after the last shellfish is sold, therefore the answer is Choice C.

45. B: Cold food items, such as meats, should be at least 41°F or cooler. Choices A, C, and D are made-up answers.

Answer Explanations #1

46. B: For fish that will be used in raw dishes, the food service operation must have documents on file that note how the fish were frozen and stored. Although Choices A, C, and D may be good information, it isn't required for the documentation.

47. B: The nine major food allergens are milk, eggs, fish, crustacean shellfish, tree nuts, wheat, peanuts, soybeans, and sesame. Choice A is incorrect and is a list of foods that have a higher likelihood of causing foodborne illness. Choice C is incorrect, and only three of the items—dairy, nuts, and seeds—are associated with the nine major food allergens. Choice D is incorrect because these are specific foods that fall into eight of the broader categories that make up the nine major food allergens.

48. B: The proper way to clean a surface includes removing any stuck-on materials, scrubbing the surface thoroughly with soapy water, rinsing with clean water, and drying the surface with a clean cloth. Wiping the surface, Choice A, may not be sufficient to remove stuck-on materials, and the surface should be scrubbed. While some surfaces can be left to air dry, Choice C, drying the wet surface with a clean towel can help prevent spills and accidents. Choice D omits the step of rinsing the surface.

49. D: Choice D is correct because the only choice provided that is not a TCS food is uncooked beans. The other choices are all TCS foods.

50. A: Insulation and other physical hazards can cause physical contamination. Choice B is incorrect because contamination with *Shigella* bacteria is considered biological contamination. Choice C is incorrect because contamination with mercury is considered chemical contamination. Choice D is incorrect because contamination with ciguatoxin is considered chemical contamination from a naturally occurring chemical.

51. A: Because the illness, toxoplasmosis, is caused by the pathogen itself—in this case the *Toxoplasma gondii* parasitic protozoan—it is classified as a foodborne infection. Choice B is incorrect because a physical hazard is an object that is present in food—such as bone fragments or staples. If classifying the *Toxoplasma gondii* parasite as a type of hazard, it would fall into the category of biological hazards rather than physical hazards. Choice C is incorrect because toxoplasmosis is common and is the second leading cause of foodborne illness-related deaths in the US. Choice D is incorrect because, although toxoplasmosis is more prevalent in some countries than in others, it is by no means isolated to developing countries, so this would not be the best classification for it.

52. A: Choice A is correct as it is a sanitary personal hygiene practice to wash hands in warm soapy water for at least twenty seconds. Choice B is incorrect because jewelry on the hands should be avoided. Choice C is incorrect because hair should be contained by a hair restraint when preparing food. Choice D is incorrect because employees should keep personal food away from where food is being prepared for customers.

53. B: Even minuscule amounts of food allergens can affect allergic individuals, so food contact surfaces on equipment and utensils must be free of any previous contamination from food allergens. Choice A is incorrect because rinsing food is not recommended for removing food allergens. Choice C is incorrect because pasteurizing and cooking are not effective means of protection against allergic reactions. Choice D is incorrect because any contact of a food allergen with food that will be consumed by an allergic individual could cause a reaction. Food that has had any contact with a food allergen to which the customer is allergic should not be served to that customer.

54. C: Breathing difficulty—along with itching, swelling, dizziness, and gastrointestinal symptoms—is a common symptom of a food allergy reaction. Choice A, jaundice, is yellowing of the eyes and skin, which is not commonly associated with food allergies but rather with liver dysfunction related to hepatitis infections. Choice B, tingling around the mouth, is not commonly associated with food allergies but rather with poisoning due to a fish toxin. Choice D, weight loss, is not commonly associated with food allergies but can be a result of other persistent foodborne illnesses, such as cyclosporiasis.

55. B: A cross connection is a physical link between clean and dirty water. There should be space between the faucet and the sink edge and between the drain and the floor, Choices A and C, as these help prevent cross connections. There is no contamination risk in a faucet that doubles as a sprayer, as in Choice D.

56. A: Backflow occurs when dirty water travels back up a hose or other connection and contaminates a clean water source, such as a faucet. Choices B and C are made-up answers involving ventilation, and Choice D is a made-up answer because fly fans generally prevent bugs from being able to enter a building.

57. B: The building's ventilation system creates adequate airflow, which helps to prevent the build-up of heat, smoke, steam, and condensation. The HVAC system, Choice A, includes heating and air conditioning. Choice C is the water control system, and Choice D is the system that provides electricity

58. C: Egg dishes that will be hot-held for service must be cooked to a minimum internal temperature of 155°F for 17 seconds. The other choices are incorrect because the temperatures they state are either too low or too high.

59. C: The only foods that are safe for reserve include unopened condiment packets like ketchup, or unopened, single-serve foods like crackers.

60. C: Michelle did everything right with this delivery, except that the shellfish did not cool to at least 41 degrees within four hours. The fish was delivered at the right temperature, Choice A, and was put into the refrigerator quickly, Choice B. The time on the truck, Choice D, does not matter as much as the delivery temperature and getting the seafood cooled further within four hours.

61. B: A sign that food has thawed and refrozen is ice crystals on the packaging. Choice A is incorrect because thawing and refreezing will not necessarily break the packaging. Choice C is incorrect as the packaging being bloated can signify that the food is no longer good, not that the food has thawed and refrozen. Choice D is incorrect as food being discolored does not indicate if it has thawed and refrozen.

62. B: Bacteria like *Clostridium perfringens*, *Clostridium botulinum*, and *Staphylococcus aureus* cause illness by creating toxins. Choice A is incorrect because bacteria can grow outside of the danger zone but do so more slowly. Choice C is incorrect because some bacteria can survive cooking; for example, they sometimes form heat-resistant spores. Choice D is incorrect because bacterial infections usually have an incubation period between the time of infection and the appearance of symptoms, usually between 12 and 72 hours.

63. C: Choices A, B, and D all outline approved methods for speeding the cooling of large batches of food. Choice C is incorrect because proofing cabinets are typically used to hold food at temperatures between 70°F and 115°F to create the optimum circumstances for dough to rise, and therefore they are not useful in cooling food to below 41°F.

64. C: When cooling hot food, the temperature of the food must decrease from 135°F to 70°F within 2 hours and from 135°F to 41°F within 6 hours, allowing for a maximum of 6 hours in the danger zone.

Choices A and B are acceptable amounts of time for the lasagna to remain in the danger zone temperature range, but they are not the maximum. Choice D is incorrect because 8 hours is too long for the lasagna to remain in the danger zone and still be safe to consume.

65. A: Choice A is correct because it states the correct process for preparing produce. Produce should be inspected, and rotten or damaged produce should be removed before being rinsed and cut. This includes produce that has a skin or rind that is not eaten. Choice B is incorrect as it states that produce that has skin that is not eaten does not need to be rinsed. Choice C is incorrect because it states that produce is rinsed before being checked for any rotten produce or damaged pieces. Choice D is incorrect as it states that produce should be washed with soap, which is not necessary.

66. B: Handwashing stations should be in or near restrooms, food prep areas, food service areas, and dishwashing areas. They are not necessary in or near dining rooms, Choices A and C, nor in bar areas, Choice D, though it is a good idea to have one in or near the bar as a best practice.

67. D: Norovirus is the primary cause of foodborne illness in the US. Choices A, B, and C are incorrect because *E. coli*, *Salmonella*, and mold are less common causes of foodborne illness in the US.

68. C: Hepatitis A is a virus. Choices A, B, and D are incorrect because disease-causing bacteria, parasites, and fungi are pathogens that are classified separately from viruses.

69. D: Choice D is correct because the minimum internal temperature that raw poultry and fowl must be cooked to is 165°F. The other choices are incorrect because the temperatures that they state are too low.

70. C: The standard shelf life for refrigerated ready-to-eat TCS foods is seven days, as long as none of their ingredients will expire before this time. If an ingredient will expire before the seven-day period, then the expiration date is the date said ingredient will expire.

71. B: Too much copper in food is considered a chemical hazard due to a toxic element, so when it causes illness, the illness is categorized as a foodborne intoxication. Choice A is incorrect because foodborne infections are illnesses caused directly by pathogens. Choice C is incorrect because copper that is leached into food in microscopic amounts is not considered a physical hazard. Fragments of metal, such as bits from a metal can, or other metal objects like jewelry that could cause physical injuries—including cuts, infections, or broken teeth—would be considered physical hazards, as would other nonfood objects like plastic, artificial fingernails, bone, or glass. Choice D is incorrect because Scromboid poisoning is a distinct foodborne intoxication that is caused by toxic levels of histamine in fish.

72. B: Choice B states 145°F as the correct minimum internal temperature at which a pork roast must be cooked. It must reach this temperature for at least four minutes. Choice A is incorrect because the temperature is too low. Choices C and D are incorrect as the temperatures given are higher than the minimum.

73. A: A purchase order or receipt is necessary to ensure that the correct amount of the correct products is included in the delivery. In addition, a thermometer is used to ensure that the meats are delivered at the appropriate temperature. A scale, Choice B, is usually used for weighing produce, and a tape measure, Choices C and D, is not generally needed for food deliveries.

74. A: Ready-to-eat foods should always be stored above raw foods to prevent raw foods from dripping contaminants into the ready-to-eat foods. Choice B is incorrect because foods should be kept below 41°F and above 135°F. Choice C is incorrect because foods should be used first in, first out, so that the oldest food is used before expiration. Choice D is incorrect because raw meat may be stored in the same cooler as produce as long as it is stored beneath.

75. C: Because this is a new supplier, Manuel should not permit a key drop delivery. Key drop deliveries are reserved for well-respected, trusted suppliers. It is not always required for management or staff to be onsite to receive a delivery, Choices A and B, if the supplier has been sufficiently vetted. Shipments should never be left unattended outside the restaurant, Choice D.

76. D: The intensity, or brightness, of lighting is measured in foot-candles, also called lux. Lumens, Choice A, are used to quantify the amount of visible light. Choice B refers to the measure of the total perceived light output, and Choice C is an electrical unit of measurement.

77. C: Choice C is correct; both chemical and high-temperature dishwashers can clean and sanitize dishes. Choice A is incorrect as not all dishes can go in dishwashers, regardless of the dishwasher being in the home or a professional establishment. Choice B is incorrect as high-temperature dishwashers can wash glassware. Choice D is incorrect as both the dishwasher and three-compartment sink can be used for cleaning and sanitizing dishes.

78. B: Choice B is correct as hot-held food should be held at 135°F and above. The other choices, while above 135°F, are not fully accurate since the minimum temperature hot-held food can be held at is 135°F. The other choices' minimum temperatures are all greater than 135°F.

79. D: cGMPs, or current Good Manufacturing Practices, are established by the Food and Drug Administration (FDA). While the USDA, Choice A, is a federal regulatory agency, it does not set forth the cGMPs. The CDFA, Choice B, is the California Department of Food and Agriculture, a regulatory agency in California (as opposed to a federal agency). The NIH, Choice C, is the National Institute of Health, a research and disease prevention agency that is not involved in food safety.

80. C: Choice C is correct because the only non-TCS food that is listed is uncooked rice. This means that the uncooked rice can be stored at temperatures above 40°F without worry of the food being unsafe for consumption. Choice A is incorrect. Cheese is a TCS food that is being held in the temperature danger zone, meaning that it is not safe for consumption. Choice B is incorrect; the chicken seems to have been thawed and refrozen, which makes the food unsafe. Choice D is incorrect since the rice is safe to be kept.

81. C: Training employees on proper hand washing and its importance for preventing foodborne illnesses is one way to address one of the five most common foodborne illness risk factors—poor personal hygiene. Choice A would be an ineffective way to check for toxin or pathogen contamination because the vast majority of toxins and pathogens are microscopic—they are not visible without a microscope. Choice B would not address any of the five foodborne illness risk factors since spreading pathogens from person to person among customers is not one of the foodborne illness risk factors. While Choice D does address a potential risk of physical contamination and is an advisable step, physical contamination through equipment such as thermometers is not considered one of the five most common foodborne illness risk factors.

82. A: Choice A states that egg yolks and whites being firm is one way, besides temperature, to know that eggs are fully cooked. Choice B is incorrect as not just the whites should be firm but also the yolks.

Answer Explanations #1

Choice C is incorrect as not just the yolks must be firm but also the whites. Choice D is incorrect because the white and the yolks must be firm, not partially firm.

83. A: Old food should be replaced with an entirely new container of food, which should be covered as it is brought from the kitchen into the self-serve area. Choice B is incorrect since new food should not be mixed in with old food. Choice C is incorrect as the new container of food should be covered as it is brought from the kitchen to the self-serve area. Choice D is incorrect as the entire container should be replaced, not just the food that it contains.

84. A: Wire shelving is best for coolers and freezers because it allows for proper airflow, which ensures the products maintain a suitable temperature. While glass and metal, Choices B and D, can help to prevent contamination from any leaks or spills, they also restrict the airflow, which impacts the equipment's ability to properly maintain temperatures. Plastics, Choice C, are not suitable as they are porous and can harbor bacteria as well as prevent proper airflow.

85. D: Outdoor garbage containers should be enclosed to prevent pests. Improved lighting can help with safety concerns, Choice A, and outside areas should have suitable drainage, Choice B, but these will not keep out pests like racoons. Garbage should never be allowed to collect inside, Choice B, and should be taken outside as often as necessary.

86. D: Hot foods should be delivered at a temperature of at least 135°F or higher. Choices A, B, and C are all incorrect.

87. C: Manufacturing procedures should be reviewed and updated routinely. Choice A, products, refers to both the raw ingredients and the final product, which should all meet quality standards. Choice B, processes, refers to the actual manufacturing, whereas the procedure is the guideline or standard for how the process is done. Choice D, people, refers to the training and specific responsibilities of the employees involved in the manufacturing process.

88. D: Choice D is correct because glassware should never be stacked. Stacking glassware allows it to come into contact with other glasses, which increases the likelihood of breakage due to stress. Also, many units will break if the stack is dropped. Choices A, B, and C all reflect best practices when working with glass.

89. A: Plumbing, lighting, ventilation, and waste disposal are all part of a building's systems and utilities. Floors, walls, ceilings, windows, and doors, Choice B, are considered part of the facilities and equipment. Choice C is part of the equipment used in a facility, and Choice D contains a combination of all these things.

90. D: All cooked vegetables are TCS foods. Choice A is incorrect because pasta and other dry grains only become TCS foods when cooked. Choice B is incorrect because raw bell peppers are not a TCS food. Choice C is incorrect because melons only become a TCS food after being cut.

Practice Test #2

1. Stirring reheated foods has which of the following safety benefits?
    a. Makes the food's texture consistent
    b. Ensures the food's internal temperature is even
    c. Helps the food reach safe temperature more quickly
    d. Allows the food to be set out for a longer time

2. When a food employee must be restricted due to illness, this means that the employee:
    a. Is permitted to work but must not handle food or handle clean surfaces that will contact food
    b. Is not permitted to work
    c. Is only permitted to work while using single-use gloves appropriately
    d. Is permitted to work but must be carefully supervised to ensure that they are no longer symptomatic

3. Suzette is having new lighting installed in her restaurant. Where should she look to find out the lighting requirements for the kitchen and food prep areas?
    a. FDA
    b. USDA
    c. Local ordinances
    d. County building codes

4. A container of potato salad was prepared on March 2 by the manufacturer. Its expiration date is March 16. The restaurant opened the container on March 4. On which date should any leftover potato salad be discarded?
    a. March 16
    b. March 11
    c. March 9
    d. March 7

5. What is one important step for proper hand washing for food handlers?
    a. Wash hands in any sink large enough to allow for the hands and arms to be thoroughly washed.
    b. Remove jewelry, such as medical bracelets, while washing and replace only after completely drying the hands.
    c. Spend at least 20 seconds on the hand washing process.
    d. If washing with soap and water is not possible, use a hand sanitizer that meets FDA guidelines.

6. If a food thermometer shows both the Fahrenheit and Celsius scales, which best represents the accuracy required to ensure that temperature measurements are adequate?
    a. +/−1°C
    b. +/−2°C
    c. +/−1°F
    d. +/−2°F

Practice Test #2

7. Which of the following is true of the differences between the three safe ways to thaw frozen foods?
   a. The microwave is the safest way to thaw food that has been frozen.
   b. Food should be cooked immediately after being thawed in the microwave or cold water. If food is thawed in the fridge, it does not have to be cooked immediately after thawing.
   c. Thawing food in the refrigerator can potentially allow for more bacteria to grow in comparison to thawing food in the microwave or submerging it in a sink filled with cold water.
   d. Thawing food in the sink is the most time-consuming option.

8. Which of the actions below would be most helpful in case of a suspected foodborne illness outbreak?
   a. Have all staff members leave the premises prior to the arrival of investigators from the regulatory entity so that they are not in the way.
   b. Discard unneeded food items to prevent them from accidentally being used later after having been left in the danger zone during the investigation.
   c. Respond to investigators' questions with yes or no answers or as succinctly as possible, avoiding details that may detract from the matter at hand.
   d. Enlist the cooperation of food employees by explaining that open and honest communication with investigators will help protect customers.

9. Which of these reportable symptoms should be reported to the regulatory authority?
   a. Jaundice
   b. Lesion with pus that is not properly covered
   c. Vomiting
   d. Sore throat with fever

10. What temperature should milk be received at?
    a. 41°F
    b. 32°F
    c. 45°F
    d. 38°F

11. Which of the following is true regarding the shelf life of condiments?
    a. Ketchup can remain unrefrigerated for up to six months.
    b. Condiments should be monitored as their shelf lives are different depending on different factors.
    c. Mayonnaise is an example of a condiment that is shelf-stable.
    d. Condiments last longest when they've been opened and are refrigerated.

12. Occasionally there is a problem with the manufacturing process that results in items needing to be destroyed or returned to the manufacturer. What is the term for this?
    a. Rejection
    b. Recall
    c. Return
    d. Rescind

13. In the United States, about how many people become sick with foodborne illnesses each year?
    a. 128,000
    b. One in six
    c. Three thousand
    d. 10%

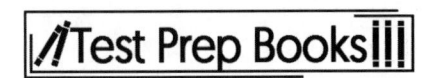

14. Which of the following situations is NOT required to be reported to a government authority?
    a. Employee slices their thumb with a knife
    b. Customer diagnosed with hepatitis A
    c. Undiagnosed employee with diarrhea
    d. Customer illness due to shellfish allergy

15. At what air temperature range should whole fruits and vegetables be stored?
    a. 50 – 60°F
    b. 60 – 70°F
    c. 70 – 80°F
    d. 80 – 90°F

16. What is the best way to handle food becoming contaminated in a self-serve area?
    a. Heat the food in order to kill any bacteria or viruses that could be within the food.
    b. Depending on what contaminated the food, there are different responses. If the food is contaminated through improper use of dishes or flatware, the food does not need to be replaced.
    c. Remove the contaminated container of food and replace it with an entirely new container of food, which is covered as it is brought out from the kitchen.
    d. Place new food atop the older, contaminated food in order to encourage the new food to be consumed first by customers.

17. Which of these is one of the five foodborne illness risk factors and possible to mitigate by incorporating specific instructions into a written recipe?
    a. Inadequate cooking
    b. Food from unsafe sources
    c. Physical hazards
    d. Paralytic shellfish poisoning

18. A hamburger is contaminated during cooking due to the fuel used in the grill. What type of contaminant is this?
    a. Sanitizing contaminant
    b. Environmental contaminant
    c. Chemical contaminant
    d. Heat-induced contaminant

19. What is the best way to keep ice safe and sanitary?
    a. Store the ice scoop on a shelf inside the machine
    b. Wash hands before scooping ice
    c. Use drinkable water to make ice
    d. Use a cup instead of an ice scoop

20. What is the minimum internal temperature that ground turkey must be cooked to?
    a. 165°F
    b. 160°F
    c. 155°F
    d. 150°F

21. Which of the following foods should NOT be rinsed?
    a. Bean sprouts
    b. Romaine lettuce
    c. Canned beans
    d. Raw chicken

22. A green bean casserole has been cooling since 12:00 PM; at 6:00 PM the casserole had an internal temperature of 42°F. What should be done with the casserole?
    a. It should go in the fridge.
    b. It should be discarded.
    c. It should be reheated and then cooled again.
    d. It should be served.

23. Food that will be reheated for hot holding must be reheated to at least 165°F for how many seconds?
    a. 5 seconds
    b. 10 seconds
    c. 15 seconds
    d. 20 seconds

24. Which of the following methods can be used to ensure that a thin chicken breast is cooked to a safe temperature?
    a. Check the inside of the chicken to ensure there is no pink color left.
    b. Insert a thermocouple thermometer into the chicken breast.
    c. Check the temperature of the chicken with an infrared thermometer.
    d. Insert the tip of the stem of a bimetallic dial thermometer into the chicken.

25. Allan is receiving a delivery and has set aside one item that must be rejected. He notified the driver and provided a description of the problem. He also made note of the incident in the store's delivery log. What did Allan forget to do?
    a. Take a picture of the rejected item.
    b. Call and notify his manager of the rejected item.
    c. File a formal complaint with the supplier.
    d. Get a signed receipt from the driver.

26. Danielle has a large stack of dishes to wash after the dinner rush as her restaurant. Which items should be washed in the dishwasher, and which items should be washed by hand?
    a. Utensils, plates, and the slicer blades by hand; pots and frying pans in the dishwasher
    b. Utensils, plates, and the slicer blades in the dishwasher; pots and frying pans by hand
    c. Utensils, plates, and frying pans in the dishwasher; slicer blades and pots by hand
    d. Utensils and plates in the dishwasher; frying pans, pots, and slicer blades by hand

27. The dishwashing machine at The Rose Café does not seem to be getting the dishes clean. Albert has inspected the machine and discovered that the high temperature thermometer is showing the water temperature to be 160°F. What should he do?
    a. Call the maintenance department; the temperature is too low.
    b. Call the maintenance department; the temperature is too high.
    c. Nothing; that temperature is within the allowable range.
    d. Even though the temperature is within the allowable range, he should call the maintenance department to see why the machine isn't working well.

28. Which of the following is NOT something to inspect when performing a daily check of the dishwashing machine?
    a. The spray nozzles are clean.
    b. Machine is free of food debris.
    c. There are no mineral deposits.
    d. The racks are stable and firmly attached.

29. What is one way to reduce the risk of physical contamination of food?
    a. Store food at or below 40°F, or at or above 140°F
    b. Avoid using sulfiting agents on raw fruits and vegetables
    c. Pay attention to seafood advisories and reject potentially affected deliveries
    d. Avoid wearing jewelry on the hands and arms when working with food

30. Which of the following is true regarding handwashing practices for professionals in a restaurant setting?
    a. Employees should wash their hands for at least fifteen seconds in soapy water.
    b. If employees wash their hands well, it is okay for them to wear jewelry on their hands.
    c. Warm, soapy water should be used while employees wash their hands for at least twenty seconds.
    d. Only employees making the food should be careful about hand hygiene.

31. Many types of seafood have the potential to cause poisoning due to fish or shellfish toxins, but some carry such high risks that it is recommended they be avoided completely. To which of the following does this recommendation apply?
    a. Mackerel
    b. Oysters
    c. Whelks
    d. Pufferfish

32. What is the minimum internal temperature that a lamb steak must be cooked to for safe consumption?
    a. 160°F
    b. 155°F
    c. 150°F
    d. 145°F

33. What is a vacuum breaker?
    a. A reverse fan used to keep bugs and pets out
    b. A stove fan that draws air up and away from the food prep area
    c. A valve that prevents water from traveling back up a hose
    d. An electrical fuse that prevents malfunction in the ventilation system

34. When using heat sanitization, the surface must reach what temperature and the surface must be soaked for how long for the sanitization to be effective?
    a. 171°F for at least thirty seconds
    b. 165°F for at least thirty seconds
    c. 180°F for at least sixty seconds
    d. 176°F for at least sixty seconds

35. Bare hands should not be used when handling food that will not be further cooked to safe temperatures before being served UNLESS:
    a. The food handler has a latex allergy.
    b. The food handler is washing fruits and vegetables.
    c. The food is not considered a time/temperature control for safety (TCS) food.
    d. The food handler washes their hands immediately before handling the food.

36. When inspecting for possible entry points for pests, which of the following areas should be checked?
    a. Cracks, gaps, and holes in walls, floors, ceilings, and window and door surrounds
    b. Openings between shelves in dry storage
    c. Underneath equipment, counters, and tables
    d. Inside cooler and freezer units and in and around stoves and ovens

37. How should single-use gloves be used?
    a. Use gloves when hands may be contaminated to avoid contaminating food.
    b. Ensure the gloves do not tear by rolling them up before putting them on.
    c. When switching tasks frequently, use gloves if there will not be time for frequent hand washing.
    d. Food handlers with artificial fingernails or nail polish should use single-use gloves for all food contact.

38. How often should floors beneath floor-mounted equipment be cleaned and inspected?
    a. Every time the rest of the floor is cleaned
    b. Every two days
    c. Weekly
    d. At the end of each shift

39. How many reportable diagnoses are there?
    a. Four
    b. Five
    c. Six
    d. Seven

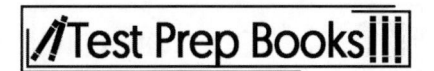

40. In an HACCP plan, if one identified critical control point is the step when prepared food is held in a warming oven before being sold, the minimum holding temperature of 135°F during this step is referred to as:
    a. A critical limit
    b. A verification procedure
    c. The danger zone
    d. A control measure

41. When thawing meat under running water, the water should be what temperature or below?
    a. 32°F
    b. 50°F
    c. 65°F
    d. 70°F

42. Which of the following frozen foods can be thawed during the cooking process?
    a. Hamburger patties
    b. Fish
    c. Pork
    d. Shellfish

43. What temperature should pastas, including baked dishes like lasagna, be cooked to?
    a. 135°F
    b. 140°F
    c. 155°F
    d. 165°F

44. Tabletop and countertop equipment should be mounted at least how high above the table or countertop?
    a. 5 inches
    b. 15 centimeters
    c. 6 inches
    d. 10 centimeters

45. What is true of the use of open condiments in a restaurant setting?
    a. Open condiments are not a contamination risk in terms of foodborne illnesses.
    b. All condiments need to be refrigerated during non-business hours.
    c. Most condiments have a longer shelf-life when left unrefrigerated.
    d. Condiments should be fully replaced when gone, not refilled into the older container.

46. What is the proper order for shelving items in refrigerators and coolers, from top to bottom?
    a. Produce and ready-to-eat foods, whole cuts of pork and beef, seafoods, ground meats and fish, poultry
    b. Seafoods, produce and ready-to-eat foods, whole cuts of pork and beef, poultry, ground meats and fish
    c. Produce and ready-to-eat foods, seafoods, whole cuts of pork and beef, ground meats and fish, poultry
    d. Seafoods, produce and ready-to-eat foods, poultry, whole cuts of pork and beef, ground meats and fish

47. What is the best way to ensure that equipment remains in good working order?
    a. Notifying management immediately of any issues
    b. Unplugging the equipment when it malfunctions
    c. Moving broken and/or nonworking equipment out of the food preparation areas
    d. Establishing a regular maintenance schedule

48. Air curtains or fly fans are best used to prevent pests in what areas of the building?
    a. External doorways
    b. Swinging doors between the kitchen and dining room
    c. Attic and basement spaces
    d. Between the kitchen and storage areas

49. What are the proper temperatures and time frames for cooling food?
    a. Hot foods should reach 70°F within two hours and 41°F within the next four hours.
    b. Hot foods should reach 65°F within two hours and 41°F within four hours total.
    c. Hot foods should reach 70°F within two hours and 41°F within four hours total.
    d. Hot foods should reach 70°F within two hours and 38°F within four hours total.

50. To allow for proper cleaning and maintenance, how high off the floor should floor-mounted equipment be?
    a. 5 centimeters
    b. 10 centimeters
    c. 15 centimeters
    d. 20 centimeters

51. Which of the following is true of sanitary practices that should be requested of customers in a self-serve area?
    a. Flatware can be grabbed by its head, so long as customers are encouraged to wash their hands.
    b. Plates can be reused by customers, so long as the self-serve utensils do not touch the plate.
    c. Signs should be posted that encourage customers to follow various sanitary practices.
    d. Sneeze guards should be used during cold and flu seasons and are not necessary during other times of the year.

52. Which of the following should be part of a regular building maintenance schedule?
    a. Cleaning and inspecting food prep equipment, disposing of old stock, and regular pest control
    b. Rotating foods according to freshness dates, cleaning and sanitizing equipment, and inspecting for damage
    c. Cleaning, inspecting building systems, pest control measures, and caring for outdoor areas
    d. Ensuring equipment is functioning properly and having regular inspections by certified technicians

53. What is the longest that ready-to-eat TCS foods can be safely stored?
    a. Three days
    b. Five days
    c. Seven days
    d. Ten days

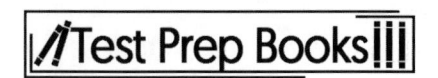

54. Cold ready-to-eat TCS foods must maintain what temperature?
    a. 45°F or lower
    b. 35°F or lower
    c. 38°F or lower
    d. 41°F or lower

55. What type of thermometer measures only surface temperatures?
    a. Thermocouple thermometer
    b. Thermistor thermometer
    c. Bimetallic stemmed (dial) thermometer
    d. Infrared thermometer

56. Which of the following is NOT true of employees working in a self-serve area?
    a. There should be employees who monitor the temperatures of the foods to ensure they stay out of the temperature danger zone.
    b. When replacing the food, employees should bring a full container of food to replace the old container.
    c. Employees should follow standard sanitary practices including having clean hands and a clean uniform.
    d. Employees must avoid working if they have had diarrhea or vomiting in the past thirty-six hours.

57. Which of the following is NOT a step in properly washing dishes using a dishwashing machine?
    a. Scrape away any food debris before loading.
    b. Load so that the spray will hit all the dish surfaces.
    c. Check that the machine is not overloaded.
    d. Towel-dry dishes and utensils after the wash and rinse cycles have finished.

58. What is the proper process to safely thaw food in a sink?
    a. The food can be removed from the packaging and submerged in cold water in a sink other than the three-compartment sink. The water should be replaced every half hour or so to keep the water cold.
    b. The sealed food should be placed in a sink of cold water. This should not be done in the three-compartment sink. The water should be replaced every hour or so to keep the water cold.
    c. The food should be sealed and placed in a sink of cold water. This should not be done in the three-compartment sink. The water should be replaced every half hour or so to keep the water cold.
    d. The food should be sealed and submerged in cold water in whatever sink is available. The water should be replaced every half hour or so in order to keep the water cold.

59. Which of the following is true regarding a Hazard Analysis Critical Control Point (HACCP) plan?
    a. It is a comprehensive approach to preventing foodborne illnesses in a food establishment.
    b. It is required for all food establishments.
    c. It assists in achieving active managerial control of foodborne illness risk factors.
    d. It should include control measures to address any potential hazards identified at a food establishment.

Practice Test #2

60. What should be used when hand washing very large items?
   a. A single large basin sink
   b. A two-basin sink
   c. A three-basin sink
   d. Wet cloths and sanitizing spray

61. What are the proper steps for cleaning and sanitizing equipment?
   a. Shut off and unplug, keep on removable parts, wipe off anything that remains on the surface, wash with a cleanser, rinse, use sanitizing solution, air dry
   b. Shut off and unplug, clean and sanitize removable parts in either a three-compartment sink or dishwasher, wipe off anything that remains on the surface, wash with a cleanser, rinse, use sanitizing solution, wipe dry
   c. Shut off and unplug, clean and sanitize removable parts in either a three-compartment sink or dishwasher, wipe off anything that remains on the surface, wash with a cleanser, rinse, use sanitizing solution, air dry
   d. Shut off, clean and sanitize removable parts in either a three-compartment sink or dishwasher, wipe off anything that remains on the surface, wash with a cleanser, rinse, use sanitizing solution, air dry

62. Gloria has received a shipment that includes a few new items for which the restaurant does not yet have regular storage space. Gloria decides to put the items into several large, airtight totes that she labels with the items, delivery date and supplier information. She stacks them out of the way in the employee locker room until space can be made in the pantry. The next day, the health inspector arrives and writes up a violation for the items. What did Gloria do wrong?
   a. She did not include the expiration dates on the totes.
   b. She did not store the totes in a suitable, food-specific location.
   c. The totes created a fire hazard by being in the employee's locker room.
   d. Items stored in non-food designated areas can only be stored for twenty-four hours.

63. Raul is inspecting a food shipment and notices that some of the packages on a particular pallet are chewed or torn. He suspects there might be a pest infestation with that pallet. What should he do?
   a. Accept the other parts of the shipment and tell the driver he won't accept that pallet.
   b. Call his manager.
   c. Leave the pallet on the truck and call the pest control company.
   d. Refuse the shipment.

64. What is a clock used for when handwashing dishes?
   a. To time how long the dishes remain in the sanitizing solution
   b. To help determine when more hot water needs to be added to the soapy water mixture
   c. To know how long to rinse each dish
   d. To time how long the dishes are in contact with the soapy water

65. Which symptoms should employees NOT have experienced for at least twenty-four hours before handling food?
   a. An infected cut on the hand
   b. Vomiting or diarrhea
   c. A sore throat and fever
   d. Jaundice

66. Which areas of the building must have potable water?
    a. Food prep areas, outside hoses, and waste areas
    b. Cleaning and sanitizing areas, external building faucets, and the bar area
    c. Handwashing stations, hostess and server stations, and the bar area
    d. Food prep areas, cleaning and sanitizing areas, and handwashing stations

67. Which of the following is an appropriate way to handle glassware to keep its structural integrity?
    a. Glassware can be allowed to soak in the cleaning compartment of the sink.
    b. If a glass has ice in it, it can be dumped out and immediately placed in the dishwasher.
    c. Glassware can be stacked atop one another, so long as they are handled with care when transporting the stack.
    d. Avoid allowing glassware to come into contact with other glassware and items.

68. If a nursing home food employee attended an event on Thursday evening where they were served food implicated in a disease outbreak, under what circumstances would restriction be necessary for the food employee reporting to work on Monday morning with no unusual symptoms?
    a. The outbreak was due to norovirus.
    b. The outbreak was due to ciguatera.
    c. The outbreak was due to hepatitis A.
    d. The outbreak was due to *Listeria monocytogenes*.

69. Which of the following best describes the concept of active managerial control?
    a. The food safety manager proactively sets up and monitors routine procedures for controlling foodborne illness risk factors.
    b. The food safety manager rotates among all food employee positions on a regular basis to identify and evaluate hazards associated with each role.
    c. The food safety manager responds to potential foodborne illness outbreaks by closing the facility, notifying the regulatory authority, and cooperating in all investigation efforts.
    d. The food safety manager uses appropriate temperature measuring devices, such as food and equipment thermometers, to ensure food is held at safe temperatures and sufficiently cooked.

70. Under which of the following conditions should a food thermometer be calibrated?
    a. After the food handler has had an interruption, such as taking a phone call
    b. When the thermometer has been left in an oven during cooking
    c. After a time–temperature indicator has changed color
    d. When the thermometer has been used on frozen food and then on hot food

71. Which of the following is a reason that a food service operation will require a variance?
    a. Serving specialty meat like bison
    b. Selling international foods
    c. Serving produce from small, local farms
    d. Packaging fresh juice to sell later

72. Hot-held food must stay at a minimum of 135°F, however, above which temperature is preferable?
    a. 138°F
    b. 141°F
    c. 145°F
    d. 155°F

73. Adriana is receiving a shipment that includes several cases of canned goods. One of the cases contains cans that appear to be dented, and a couple of the cans are missing their labels, though the delivery driver says all the cans in the case are the same product. What should Adriana do?
    a. Accept the delivery; dents are not a problem, and the contents of the cans is known.
    b. Reject the delivery due to the dents and missing labels.
    c. Accept the delivery with a certified statement from the driver regarding the products' safety.
    d. Reject the delivery due to the dents, even though the contents of the cans is known.

74. Checking shipment packaging for evidence of pests is part of which step in the pest control process?
    a. Deprivation of food and water
    b. Preventing access
    c. Treatment
    d. Inspection

75. When handwashing dishes, what temperature should the water be for the soap and water mixture?
    a. At least 100°F
    b. At least 105°F
    c. At least 110°F
    d. At least 125°F

76. If a food establishment has a written process specifying that once per week, the general manager will review the food temperature log maintained by the kitchen manager to ensure the records are adequate and that appropriate corrective actions are taken, which HACCP principle is it applying?
    a. Establish verification procedures.
    b. Establish corrective actions.
    c. Establish monitoring procedures.
    d. Establish record-keeping procedures.

77. Travis is making a batch of coleslaw on October 12 that the restaurant will begin using immediately. The mayonnaise that he is using has an expiration date of October 16. What expiration date should be put on the label?
    a. No label is needed on freshly made items.
    b. No expiration date is needed because the food is being used immediately.
    c. October 16
    d. October 19

78. If an item from a delivery needs to be rejected, what should be done with the item immediately?
    a. It should be thrown away.
    b. It should be given back to the driver.
    c. It should be separated from the rest of the delivery.
    d. It should be put away with the rest of the stock but with a note not to use.

79. In order to prevent pests from accessing food stored on shelves, how high off the floor should the shelves be lifted?
    a. 2 inches
    b. 4 inches
    c. 6 inches
    d. 8 inches

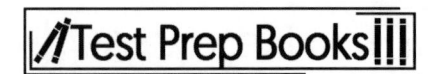

80. Sampson is emptying the trash at the end of the food prep counter. The bag is very heavy, and he noticed that there is something leaking from the bottom corner of the bag. He rests the bag on the edge of the counter so that it doesn't break open and he can tie it tightly before carrying it outside. What has he done incorrectly?
    a. There should not be liquids in the garbage that could cause a leak.
    b. He should not rest the bag on the counter.
    c. The bag should have been emptied sooner so it isn't so heavy.
    d. The bag should be tied while still in the can.

81. Allison is inspecting her building for pest entry points and has discovered two vents that are not covered. What should she use to cover these vents?
    a. Metal caps
    b. Boards
    c. Wire mesh screens
    d. Nothing; vents cannot be covered

82. Which of the following is NOT a service typically provided by a pest control company?
    a. Identify the source of the infestation.
    b. Seal cracks and other openings where pests are entering the building.
    c. Provide suggestions for prevention of future problems.
    d. Set traps and spray chemicals to help eliminate pests.

83. Cold-held food from the refrigeration is safe for up to six hours as long as:
    a. The internal temperature doesn't rise above 70°F
    b. The food is out of the sun
    c. The food is kept on a cold table
    d. The air temperature doesn't rise above 90°F

84. Which of these is the most appropriate way to cover a boil located on a food worker's finger?
    a. With an impermeable barrier, such as a cot
    b. With an impermeable barrier and a single-use glove
    c. With an impermeable barrier and a slash-resistant glove
    d. With a dry, tight-fitting bandage

85. Martin has noticed a problem with odors being detected in the dining areas of his restaurant. Which building system should be inspected to resolve this problem?
    a. Waste disposal
    b. Plumbing
    c. Food storage
    d. Ventilation

86. Which two areas should include an air gap in a properly installed sink?
    a. Between the faucet and the rim of the sink and between the drainpipe and the floor drain
    b. Between the faucet and the sink basin and between the sink cabinet and the floor drain
    c. Between the sink basin and the wall and between the sink basin and the floor
    d. Between the sink basin and any handheld sprayers and between the drain plug and the drainpipe

87. A shipment of dairy goods has arrived at Denise's restaurant, and she has been tasked with receiving the delivery. She notices that the milk cartons have several different expiration dates. Today's date is February 27. Which of the following cartons should be rejected?
    a. Two cartons dated February 26
    b. Four cartons dated March 4
    c. One carton dated March 2
    d. One carton dated March 8

88. What is the minimum internal temperature that fish must be cooked to?
    a. 145°F
    b. 150°F
    c. 155°F
    d. 160°F

89. Sarah is storing items from a recent delivery. She is transferring some of the goods from their original packaging into larger storage containers that are labeled with the brand name of each product, the delivery receipt date, and the expiration date. What is missing from Sarah's labels?
    a. The common name of the product
    b. The name of the supplier
    c. The location of the manufacturer
    d. The phone number of the supplier

90. Which of the following foods can be held safely at room temperature?
    a. Pork roast
    b. Cooked beans
    c. Potatoes
    d. Ranch dressing

Answer Explanations #2

1. B: Food should be stirred during the reheating process—especially if using a microwave oven—because this combines the warm and cold parts of the food so that it reaches a consistent safe internal temperature. Choice A is incorrect because this is not a safety benefit. Choices C and D are incorrect because the answers are not true.

2. A: Food employees who have had certain reportable symptoms or reportable diagnoses may, under specific circumstances, be permitted to work; however, they will be restricted from food handling duties. Choice B is incorrect because the correct term for not permitting an employee to work is to *exclude*, not *restrict*. Choices C and D do not represent measures that would fulfill the requirements of restricting an employee due to illness.

3. C: Local ordinances regulate the lighting requirements for food prep locations. The FDA and USDA, Choices A and B, are regulatory agencies for food safety, but they do not regulate buildings and facilities. County building codes, Choice D, determine the structure of the building for its stability and safety, but they do not regulate lighting for the food industry.

4. B: Products should be discarded within seven days of opening, which would be March 11 in this example. This is true regardless of the expiration date from the manufacturer, Choice A. Choices C and D are made-up answers.

5. C: The hand washing process, from wetting the hands and arms to drying them, should take at least 20 seconds, and at least 10–15 seconds should be spent vigorously rubbing the hands and arms with soapy water. Choice A is incorrect because only dedicated hand washing sinks should be used to wash hands. Choice B is incorrect because no jewelry, other than a plain ring, should be worn when working with food. Choice D is incorrect because hand sanitizers cannot be used as a substitute for hand washing and should only be used as an optional step after washing hands.

6. A: Food thermometers that display a Celsius scale should be accurate to +/−1°C. Choice B is incorrect because +/−2°C is a less stringent accuracy requirement and may lead to food temperature measurements that falsely indicate a food is at a safe temperature. Choice C is incorrect because, although 1°F is a smaller unit than 1°C, and this standard would require more accuracy than +/−1°C, this choice does not best represent the accuracy requirement. Choice D is incorrect because +/−2°F is less accurate than +/−1°C and is only an acceptable standard when a thermometer does not have a Celsius scale.

7. B: Choice B states a correct difference between the three safe ways to thaw frozen foods. Food thawed in the fridge does not need to be immediately cooked, while if it's thawed in the microwave or sink it does need to be cooked immediately. Choice A is incorrect. Since the microwave can potentially partially cook the food, it must be immediately cooked for safety. Choice C is incorrect as thawing food in the fridge keeps the food out of the temperature danger zone, which means that bacteria cannot grow. Choice D is incorrect because thawing food in the fridge is the most time-consuming option, not the sink.

8. D: Encouraging food employees' cooperation in the investigation of a potential foodborne illness outbreak is helpful for the investigation. Choice A may not be helpful since investigators may need information from staff members to aid in the investigation. Choice B is incorrect because no food should be thrown away, as it may be needed to help determine the cause of the outbreak. Choice C is incorrect

Answer Explanations #2

because investigators will need open communication and detailed accounts to help determine the cause of the outbreak.

9. A: When a food employee is symptomatic with jaundice—yellow eyes or skin—the regulatory authority must be notified and must provide approval before the employee returns to work. Choice B is incorrect because lesions with pus that are not properly covered are a reportable symptom that results in restricted duties, but reporting them to the regulatory authority is not necessary. Choice C, vomiting, along with diarrhea, is a reportable symptom that warrants exclusion from work but does not need to be reported to the regulatory authority. Choice D, sore throat with fever, is a reportable symptom that warrants exclusion or restriction, depending on whether the facility primarily serves a highly susceptible population (HSP), but reporting it to the regulatory authority is not necessary.

10. C: Milk should be received at an internal temperature of 45°F or cooler and should be cooled to 41°F or cooler within four hours.

11. B: Condiments should be monitored as their shelf lives are different depending on different factors such as whether they're opened, refrigerated, etc. Choice A is incorrect because ketchup can remain unrefrigerated for one year if unopened or for one month if opened. Choice C is incorrect because mayonnaise is not shelf-stable and should not be out of refrigeration for more than two hours. Choice D is incorrect because condiments' shelf lives are generally longer when they are unopened.

12. B: A product recall occurs when a problem occurred during manufacturing that was not discovered until after the product had been shipped to customers. Choices A, C, and D are made-up answers.

13. B: One in six people become sick with foodborne illnesses each year in the US. Choice A represents the number of hospitalizations each year in the US, and Choice C represents the number of deaths each year in the US. Choice D is incorrect because one in six is approximately 17% of the population, not 10%.

14. C: An employee with diarrhea is not permitted to work in food preparation, but without a doctor's diagnosis of a reportable illness, reporting this symptom is not mandatory. Thus, Choice C is correct. Choice A is incorrect because an on-the-job injury must be reported to OSHA. Choices B and D are incorrect because all customer reports of illness must be reported, regardless of cause.

15. B: Whole fruits and vegetables should be stored between 60°F and 70°F, therefore Choice B is correct.

16. C: Choice C is correct as contaminated food should be removed entirely and replaced with a new container. Choice A is incorrect because contaminated food should be removed. Some things that can contaminate food cannot be killed through heat. Choice B is incorrect because contaminated food is contaminated, regardless of the source of that contamination, and therefore should be replaced. Choice D is incorrect since older, contaminated food should be removed so it is not served.

17. A: Incorporating specific instructions for the cooking time and temperature into a written recipe will help prevent inadequate cooking. Although food from unsafe sources, Choice B, is a foodborne illness risk factor, it is one that must be addressed at the time of receiving food by ensuring food is only received from approved sources and is therefore outside the scope of a recipe card. Choices C and D are not among the five most common foodborne risk factors, and additionally, Choice D—paralytic shellfish poisoning—is due to toxins that cannot be deactivated by cooking.

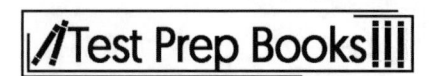

18. C: Chemicals added to foods unintentionally due to the use of chemicals throughout the production process, including things like fuel, sanitizers, and cleaning products is considered chemical contamination. Thus, Choice C is correct. Choices A and D are incorrect because these terms are not recognized as specific types of contaminants. Choice B is incorrect because environmental contaminants enter food from the ecosystem.

19. C: Only drinkable water should be used to make ice. Choice A is incorrect because ice scoops should be stored outside the machine, not inside. Choice B mentions handwashing but leaves out the additional step of glove wearing to get ice, and Choice D is incorrect because only an ice scoop should be used to get ice.

20. A: Choice A is correct as it states the correct temperature that any cut of poultry or fowl should be cooked to 165°F. The other choices are incorrect as the temperatures they state are all lower than 165°F, which would make the ground turkey unsafe to eat.

21. D: Choice D correctly states that raw chicken should not be rinsed due to contamination risks. The other choices—bean sprouts, romaine lettuce, and canned beans—are all able to be rinsed and are safe to rinse.

22. B: Since the casserole has not reached an internal temperature of 41°F or less in six hours, it should be discarded. It is too late to reheat it, and it is unsafe to serve or keep.

23. A: Foods that will be hot-held must be reheated to at least 165°F for five seconds. The other answer choices are incorrect times.

24. B: Thermocouple thermometers can be used to measure the temperature of foods of any thickness, unlike bimetallic thermometers. Choice A is incorrect because visual indicators like color are not a reliable indicator of safety. Choice C is incorrect because infrared thermometers are only able to measure surface temperatures, and internal temperatures must be measured to ensure that food is safely cooked. Choice D is incorrect because the sensor of a bimetallic stemmed (dial) thermometer is usually 2–3 inches long, from the tip of the stem to the dimple, and this entire area must be inserted into the chicken to accurately measure the temperature, not just the tip.

25. D: The delivery driver should provide a signed receipt of adjustment or credit when an item from a delivery is rejected. Choices A, B, and C are not standard procedure for handling a rejection, although different facilities may have different requirements. Managers and staff should always have a clear policy for the procedures required by their particular facility.

26. B: The dishwasher is good for washing utensils, plates, and other small items, such as the slicer blades. Large items like pots and frying pans should be washed by hand. Choices A, C, and D are incorrect.

27. A: The final sanitizing rinse temperature for most dishwashing machines should be between 165 and 180°F. Albert needs to call maintenance because 160°F is too low. Choices B, C, and D are incorrect.

28. D: There is not usually a need to check the stability of the racks in the dishwasher unless they are clearly coming loose or have fallen. However, Choices A, B, and C should be inspected daily to ensure that the machine stays in good working order.

29. D: Jewelry or parts of jewelry can accidentally fall off into food and create dangerous physical hazards for consumers through physical contamination. Choice A is incorrect because avoiding the

Answer Explanations #2

danger zone, between 40°F and 140°F, will help control biological hazards by inhibiting the growth of certain pathogens. Choice B is incorrect because sulfiting agents are chemical preservatives that are considered chemical hazards when used improperly on raw fruits and vegetables in food establishments. Choice C is incorrect because observing fish advisories is a way to reduce the risks associated with chemical contamination of seafood at its source.

30. C: Employees should wash their hands in warm soapy water for at least twenty seconds. Choice A is incorrect because the minimum time for hand washing is twenty seconds, not fifteen seconds. Choice B is incorrect because jewelry on the hands should generally be avoided, even with clean hands. Choice D is incorrect because all employees should be careful about hand hygiene, not just those making the food.

31. D: Pufferfish, also known as fugu or blowfish, carry a high risk of tetrodotoxin poisoning, which is life-threatening, and these fish should be avoided. Choice A is incorrect because, although mackerel is associated with scromboid poisoning and ciguatera, it is not as high-risk. Choice B is incorrect because, although oysters are associated with some types of shellfish poisoning, they are not as high-risk. Choice C is incorrect because, although whelks are associated with neurotoxic shellfish poisoning, this is not usually life-threatening.

32. D: Choice D is correct because the minimum internal temperature that a lamb steak must be cooked to for safe consumption is 145°F, which it much reach for a minimum of 15 seconds. The other choices all state a temperature that is above the minimum internal temperature that a lamb steak must be cooked to for safe consumption.

33. C: A vacuum breaker prevents backflow by shutting off the water supply line when the water is turned off.

34. A: Heat sanitization involves heating the surface to at least 171°F and soaking for at least thirty seconds. Choices B, C, and D are incorrect.

35. B: Food that will not be further cooked to safe temperatures is considered ready-to-eat (RTE) food and should not be handled with bare hands except when washing produce. Choice A is incorrect because a latex allergy does not exempt a food handler from wearing gloves. Nonlatex gloves should be provided, and utensils can also be used to prevent bare hand contact. Choice C is incorrect because gloves should be worn with RTE foods regardless of whether they are considered TCS foods. Choice D is incorrect because, unless the facility has the appropriate permit from a regulatory authority, bare hand contact is not acceptable even when hands have been washed.

36. A: Pests can enter a building through the tiniest of openings, including cracks, gaps, and holes in walls, floors, ceilings, and window and door surrounds. These should all be checked for and repaired regularly. While there may be evidence of pests in dry storage, Choice B, this is not usually an access point to the building, nor are the areas beneath equipment, counters, and tables, Choice C, even though there may be evidence of infestation in these places. Similarly, Choice D can provide evidence of pests, but these are not entry points.

37. D: Artificial fingernails and nail polish present a physical hazard when working with food and should be avoided or covered with single-use gloves when working with food. Choice A is incorrect because gloves should not be put on when hands are contaminated. Hands should always be washed prior to putting on gloves. Choice B is incorrect because gloves should never be rolled up while being put on.

Choice C is incorrect because, while gloves should be changed when switching tasks, hand washing would be necessary anyway since it should be done before putting gloves on.

38. A: The floor beneath all equipment should be cleaned and inspected every time the rest of the floor is cleaned. Choices *B*, *C*, and *D* are incorrect.

39. C: There are six reportable diagnoses—the illnesses caused by norovirus, hepatitis A, *Shigella*, Shiga toxin-producing *E. coli* (STEC), *Salmonella* Typhi (typhoid fever), and nontyphoidal *Salmonella*. Choices *A*, *B*, and *D* are incorrect because they do not represent the correct number of reportable diagnoses.

40. A: A critical limit in an HACCP plan is a numerical measurement, such as a temperature, used to determine whether safety standards are met at a critical control point. Choice *B*—a verification procedure—is a written description of how an HACCP plan's efficacy will be verified. Choice *C*—the danger zone—is the temperature range between 41°F and 135°F in which pathogens grow most rapidly, fostering contamination from pathogen overgrowth and toxin formation. Choice *D*—a control measure—is an action taken to eliminate or reduce the risk of a hazard at a critical control point.

41. D: When using the running water method to thaw meat, the water temperature should be 70°F or lower.

42. A: Only a few types of frozen foods, including frozen chicken that will be fried and frozen hamburger patties, can safely be used frozen and thawed during the cooking process.

43. D: Dishes that contain pasta, like lasagna, should be cooked to at least 165°F.

44. D: Table- and counter-mounted equipment should be at least 4 inches, or 10 centimeters, above the surface. Floor-mounted equipment should be 6 inches, or 15 centimeters, above the floor, Choices *B* and *C*.

45. D: Condiments should be fully replaced and not refilled into the old container. Choice *A* is incorrect as open condiments can be a contamination risk through their use and ability to go bad. Choice *B* is incorrect because not all condiments necessarily need to be refrigerated outside of business hours. Choice *C* is incorrect; when not refrigerated, condiments' shelf lives are generally shorter than when refrigerated.

46. C: Produce and ready-to-eat foods, seafoods, whole cuts of pork and beef, ground meats and fish, and poultry should be stored in this order, from top to bottom. Foods should be stored so that any drips from meats or other possible contaminants do not get on fresh or other food items. In Choices *A*, *B*, and *D*, the fresh foods and seafoods could possibly be contaminated by the meats.

47. D: A regular maintenance schedule, both in-house and with a technician, can help avoid any problems with the equipment. While it's good practice to notify management of any issues, Choice *A*, along with unplugging malfunctioning equipment, Choice *B*, these are protocol and safety precautions after a problem is evident rather than preventative measures. Moving broken equipment, Choice *C*, is not usually a feasible option.

48. A: Air curtains and fly fans are used in external doorways to prevent bugs from entering through the open doors. Choices *B*, *C*, and *D* are incorrect.

49. C: Choice *C* states the correct time frames and temperatures regarding cooling food. Hot foods should reach 75°F within two hours and 41°F within four hours total. Choice *A* is incorrect as the food

Answer Explanations #2

should reach 41°F within four hours total, not four hours following the first two hours. Choice B is incorrect because hot foods should reach 70°F within the first two hours, not 65°F. Choice D is incorrect because hot foods should reach 41°F within four hours, not 38°F.

50. C: Floor-mounted equipment should be at least 15 centimeters (6 inches) off the floor. Choices A, B, and D are incorrect.

51. C: Choice C states that signs should be posted to encourage customers to follow sanitary practices, which is correct. Choice A is incorrect since flatware should be grabbed by the handle. Choice B is incorrect because plates should not be reused in a self-serve area. Choice D is also incorrect; sneeze guards do not have a specific timeframe that they should be used within; rather they should be used generally.

52. C: Cleaning, inspecting building systems, pest control measures, and caring for outdoor areas are all part of a comprehensive maintenance schedule. Choices A, B, and D all include tasks that, while important, are not specifically part of building maintenance.

53. C: Ready-to-eat TCS foods cannot be stored for longer than seven days. Choices A, B, and D are incorrect.

54. D: Cold ready-to-eat TCS foods must maintain a temperature of 41°F or colder to prevent the growth of harmful bacteria. Choices A, B, and C are incorrect answers.

55. D: Infrared thermometers measure the surface temperatures of food and other equipment. Choices A, B, and C are types of thermometers with probes or stems that should be inserted into the food or liquid in order to register the temperature.

56. D: Employees must avoid working with food if they have had diarrhea or vomiting within the past twenty-four hours, not thirty-six hours. Choice A is incorrect because it is true that there should be employees who monitor the temperatures of foods to ensure they stay out of the temperature danger zone. Choice B is incorrect because it is true that when food is replaced, the entire container should be replaced. Choice C is incorrect because it is true that employees should follow standard sanitary practices including having clean hands and a clean uniform.

57. D: Items washed in the dishwasher should not be dried with a towel or cloth, as this could cause recontamination. Instead, allow the dishes and utensils to air dry. Choices A, B, and C are all important steps in ensuring that dishes get washed properly in the machine.

58. C: Choice C is correct as it states the proper process to safely thaw food in a sink. The food should be sealed and placed in a sink of cold water, but not in the three-compartment sink. To keep the water cold, it should be replaced every half hour or so. Choice A is incorrect because the food should not be removed from the packaging; it should be sealed. Choice B is incorrect because the water should be replaced every half hour or so, not every hour. Choice D is incorrect; the food should not be placed in any sink available since it should not be done in a three-compartment sink.

59. C: An HACCP plan is one element that a food establishment can incorporate as part of its approach to controlling foodborne illness risk factors. Choice A is incorrect because an HACCP plan is not comprehensive—it addresses particular hazards and must be used in addition to the facility's standard procedures, known as prerequisite programs. Choice B is incorrect because an HACCP plan is only required in certain facilities, not all. Choice D is incorrect because not all hazards identified at a food establishment must be addressed in an HACCP plan. Principe 1 of the HACCP system involves identifying any potential hazards and evaluating each one's likelihood of occurring and severity of the risk it could cause in order to determine which of the hazards will be addressed through control measures.

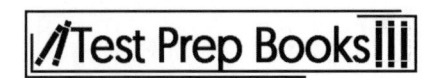

60. C: A three-basin sink is best for washing large items, as it allows a basin for washing, rinsing, and sanitizing. Single or two-basin sinks, Choices A and B, are not generally good for this as they do not allow separate areas for each step and could cause cross-contamination. Choice D is not a safe option for washing cooking utensils or dishes.

61. C: Choice C states the correct process to clean and sanitize equipment. Once the equipment is shut off and unplugged, the removable parts can be removed and washed separately. The equipment should be wiped clean of anything remaining on the surface before it is cleansed and rinsed. Then, the sanitizing solution should be used, and the equipment and its parts should be allowed to air dry. Choice A is incorrect as it states that removable parts should be washed along with the rest of the equipment. Choice B is incorrect as it states that the equipment and its parts should be wiped dry after being sanitized. Choice D is incorrect as it only states that the equipment should be shut off, not that it should be unplugged.

62. B: Food items should never be stored in non-food-specific areas, such as the locker room, no matter how well they are packaged and/or stored. Choices A, C, and D are made-up answers.

63. D: Any shipment that shows evidence of pests should be immediately refused. Accepting a partial shipment, Choice A, could be risky as there could be pests that are yet undiscovered. While the company policy may allow accepting a partial shipment, or may require that the manager be called, Choice B, the best practice is to refuse the entire shipment in the interest of caution. A pest control company generally will not inspect delivery trucks while they are on site, Choice C. It would, however, be a good idea for the delivery driver to contact his company and follow their guidelines for dealing with pests on their trucks.

64. A: Sanitizing processes that involve soaking dishes require a certain amount of time for the product to effectively kill all the germs and bacteria. A clock near the dishwashing station can be used to ensure that the dishes are in the sanitization solution for enough time. Determining whether the soapy water has cooled too much, Choice B, can be accomplished by touch or thermometer rather than by time. There is no set amount of time for rinsing dishes, Choice C. They should be rinsed until the soap is gone. Similarly, there is no set time for washing/soaping the dishes, Choice D.

65. B: Choice B is correct since it states that employees should not have had vomiting or diarrhea for twenty-four hours in order to handle food. The other choices are incorrect; while they are important symptoms to track in order to prevent illness from being passed through food, they each have their own protocol. It is not stated that they should not have occurred for twenty-four hours before handling food.

66. D: Food prep areas, cleaning and sanitizing areas, and handwashing stations must all have access to clean, potable water. It is not generally required that outside water sources, such as in Choices A and B, have potable water, though if the water being provided to the building is potable, chances are these areas will have access to that water as well. Hostess stations, Choice C, are not required to have direct access to water.

67. D: Choice D states that glassware should not be allowed to come into contact with other glassware or items, which is correct. This hurts the structural integrity of the glass and increases the risk of it breaking. Choice A is incorrect; glassware soaking in the sink can lead to glassware coming into contact

Answer Explanations #2

with one another or other dishes. Choice B is incorrect because dumping the ice out of the glass and washing it in a dishwasher can cause thermal shock due to the temperature of the glass changing too quickly. Choice C is also incorrect, as glassware should not be stacked atop one another because this can lead to glassware being dropped and broken.

68. C: Food employees who may have been exposed to the hepatitis A virus through food involved in a confirmed disease outbreak within the past 30 days should be restricted from work in a facility that serves primarily highly susceptible populations (HSPs), such as a nursing home. Choice A is incorrect because potential exposure to norovirus necessitates restriction from facilities serving primarily HSPs only if the exposure occurred within the past 48 hours. Because Thursday evening to Monday morning is longer than 48 hours, the food employee does not need to be restricted when reporting for work on Monday. Choices B and D are incorrect because potential exposure to ciguatera toxin or *Listeria monocytogenes* does not require restriction from work in a food establishment setting.

69. A: Active managerial control is a way to address foodborne illness risk factors through proactive, routine procedures. Choice B may be helpful for fulfilling principle 1 of the HACCP system—conducting a hazard analysis—but it does not describe the concept of active managerial control. Choice C is a true statement related to responding to a potential foodborne illness outbreak but does not describe the concept of active managerial control. Choice D describes a monitoring procedure that could be used as part of a food establishment's plan for active managerial control of foodborne illness risk factors, but it does not encompass the entire scope of the concept.

70. D: Food thermometers should be calibrated after extreme changes in temperature. Choices A and B are incorrect because these circumstances—an interruption in food handling tasks and being left in food while cooking—do not automatically necessitate calibrating the thermometer. Choice C is incorrect because time–temperature indicators are a type of temperature measuring device that cannot be calibrated.

71. D: Only a few things require the issuance of a variance, including food service operations that grow and sell sprouts or package fresh juice to sell at a later time.

72. B: Temperatures of 141°F or higher are preferable when keeping hot-held foods.

73. B: Products that have packaging that is dented and/or are missing labels should be rejected. A statement from the driver about the contents of the cans, Choices A, C, and D, is not sufficient.

74. B: Inspecting shipment packing for evidence of chewing or tearing, as well as inspecting for egg cases, droppings, and body parts, is part of the prevention step of pest control. Choices A and C are the second and third steps in the process. Choice D, while important, is not one of the specific steps in pest control.

75. C: For washing dishes, soapy water should be at least 110°F. Choices A, B, and D are incorrect.

76. A: In HACCP principle 6, verification procedures consist of a written description of the actions that will be taken to verify the HACCP plan's efficacy, including verifying that monitoring procedures—including maintaining a temperature log—are being performed properly. Choice B—establishing corrective actions, the fifth HACCP principle—would be applied through a written procedure for taking corrective action. In the example given, a protocol for how to respond to a food temperature measurement that does not meet the applicable critical limit, such as discarding or further heating the food, would be part of establishing corrective actions. Choice C—establishing monitoring procedures, the fourth HACCP principle—would be applied through a written procedure for how critical limits are to be monitored. In the example given, a procedure explaining that the kitchen manager is to check food temperatures on a regular basis would be part of establishing monitoring procedures. Choice D—establishing record-keeping procedures, the seventh HACCP principle—would be applied through a written procedure for how to record actions taken to fulfill prerequisite programs and HACCP plan procedures. It may include how monitoring measurements and observations, corrective actions, and verification actions are to be documented.

77. C: The product should be discarded by the earliest expiration date for any ingredient, or within seven days of preparation. In this case, the mayonnaise expires earlier than the seven days, which would be Choice D, so Choice C, October 16, is the date the leftover product should be discarded. Choices A and B are incorrect answers.

78. C: A rejected item from a delivery should be immediately separated from the rest of the delivery until it can be dealt with. The supplier should provide instructions as to whether the item should be thrown away, Choice A, or returned to the driver, Choice B. The item should never be placed with the usable supplies, Choice D.

79. C: Shelves should be raised at least 6 inches off the floor to prevent pests from accessing the stored items. Choices A, B, and D are not the correct distance.

80. B: Garbage should never come into contact with food prep areas, including counters. While it is best to pour liquid waste into a drain when suitable, there are things that leak that will end up in the garbage, Choice A. Good quality commercial garbage bags can help prevent leaks. Garbage should be taken out regularly, but there is not really a way to control how heavy a bag is, Choice C, without perhaps being wasteful by taking out half-full bags. Tying the bag while it is still in the trash can, Choice D, is not always feasible, though can be done sometimes.

81. C: Vents should be covered with wire mesh screens to keep bugs out while still allowing for suitable airflow and ventilation. Covering vents with solid materials such as metal caps or boards, Choices A and B, would restrict air flow. Vents can be covered if suitable material is used, so Choice D is incorrect.

82. B: Pest control companies do not generally perform works such as sealing cracks or other openings. They will, however, make suggestions as to how to prevent future pests, Choice C, including suggestions for sealing openings. They will also work to identify the source of the infestation, Choice A, and eliminate the pests, Choice D.

83. A: After being removed from refrigeration, cold held food is safe as long as it doesn't reach an internal temperature above 70°F. While cold tables may help keep the temperature down, they aren't the only way to do so safely. Air temperature and sunshine do not affect cold-holding standards, although they may cause the food to warm up faster.

84. B: When located on the hand or wrist, a lesion with pus, such as a boil, should be covered with an impermeable barrier and further covered by a single-use glove. Choice A is incorrect because, although an impermeable cot or other bandage should be used to cover the boil, the cot should further be covered by a single-use glove. Choice C is incorrect because the impermeable barrier (bandage, cot, etc.) should be covered by a single-use glove, which has an impermeable surface. Most slash-resistant gloves do not have an impermeable surface. Choice D is incorrect for a boil on the finger but would be appropriate for a boil located on a part of the body besides the exposed arms and hands. To be a

Answer Explanations #2

sufficient cover for a lesion on the hand, the cover would also need to be impermeable and covered by a single-use glove.

85. D: The ventilation system controls the airflow in a facility. If odors are being detected, there is likely a problem with this system. While problems in other areas, such as Choices A, B, and C, could create odors, the spread of these odors is controlled by the ventilation.

86. A: A properly installed sink will have air gaps between the faucet and the rim of the sink and between the drainpipe and the floor drain. This prevents contamination between clean water flowing from the faucet and water that is standing in the sink, even when the sink is full, and between water draining from the sink and the dirty water that may be standing on or around the floor drain. Choices B, C, and D are made-up answers.

87. A: Any item that is beyond its expiration date, even if that date is very recently passed, should be rejected. Choices B, C, and D can be accepted and should be stacked for use according to the earliest expiration date.

88. A: Choice A is correct as the minimum internal temperature that fish must be cooked to is 145°F for at least 15 seconds. The other choices are all higher than the minimum temperature that fish must be cooked to.

89. A: It is always a good practice to label items with the common name, such as "flour" or "sugar," rather than the brand name to avoid any confusion. Manufacturer and supplier information, Choices B, C, and D, are not necessary on the product labels, though the facility should have records of that information on file.

90. C: Choice C is correct as potatoes can be stored safely at room temperature. The other choices—pork roast, cooked beans, and ranch dressing—are all TCS foods and should not be stored at room temperature due to illness risks.

Dear ServSafe Test Taker,

Thank you again for purchasing this study guide for your ServSafe Test. We hope that we exceeded your expectations.

Our goal in creating this study guide was to cover all of the topics that you will see on the test. We also strove to make our practice questions as similar as possible to what you will encounter on test day. With that being said, if you found something that you feel was not up to your standards, please send us an email and let us know.

We have study guides in a wide variety of fields. If the one you are looking for isn't listed above, then try searching for it on Amazon or send us an email.

Thanks Again and Happy Testing!
Product Development Team
info@studyguideteam.com

FREE Test Taking Tips Video/DVD Offer

To better serve you, we created videos covering test taking tips that we want to give you for FREE. **These videos cover world-class tips that will help you succeed on your test.**

We just ask that you send us feedback about this product. Please let us know what you thought about it—whether good, bad, or indifferent.

To get your **FREE videos**, you can use the QR code below or email freevideos@studyguideteam.com with "Free Videos" in the subject line and the following information in the body of the email:

   a. The title of your product

   b. Your product rating on a scale of 1-5, with 5 being the highest

   c. Your feedback about the product

If you have any questions or concerns, please don't hesitate to contact us at info@studyguideteam.com.

Thank you!